Dynamic Programming in Econo

Dynamic Modeling and Econometrics in Economics and Finance

VOLUME 5

Aims and Scope

The series will place particular focus on monographs, surveys, edited volumes, conference proceedings and handbooks on:

- Nonlinear dynamic phenomena in economics and finance, including equilibrium, disequilibrium, optimizing and adaptive evolutionary points of view; nonlinear and complex dynamics in microeconomics, finance, macroeconomics and applied fields of economics.

- Econometric and statistical methods for analysis of nonlinear processes in economics and finance, including computational methods, numerical tools and software to study nonlinear dependence, asymmetries, persistence of fluctuations, multiple equilibria, chaotic and bifurcation phenomena.

- Applications linking theory and empirical analysis in areas such as macrodynamics, microdynamics, asset pricing, financial analysis and portfolio analysis, international economics, resource dynamics and environment, industrial organization and dynamics of technical change, labor economics, demographics, population dynamics, and game theory.

The target audience of this series includes researchers at universities and research and policy institutions, students at graduate institutions, and practitioners in economics, finance and international economics in private or government institutions.

Dynamic Programming in Economics

by

Cuong Le Van
French National Center for Scientific Research
University of Paris 1, France

and

Rose-Anne Dana
University of Paris-Dauphine, France

KLUWER ACADEMIC PUBLISHERS
DORDRECHT / BOSTON / LONDON

A C.I.P. Catalogue record for this book is available from the Library of Congress.

ISBN 978-1-4419-5347-6

Published by Kluwer Academic Publishers,
P.O. Box 17, 3300 AA Dordrecht, The Netherlands.

Sold and distributed in North, Central and South America
by Kluwer Academic Publishers,
101 Philip Drive, Norwell, MA 02061, U.S.A.

In all other countries, sold and distributed
by Kluwer Academic Publishers,
P.O. Box 322, 3300 AH Dordrecht, The Netherlands.

Printed on acid-free paper

Printed in the Netherlands.

Table of Contents

Preface

This book is an outgrowth of a course intended for students in the first year PhD program, and for researchers in Macroeconomic Dynamics. It can be used by students and researchers in Mathematics, as well as in Economics as. The purpose of the book is two fold: (a) to provide a rigorous, but not too complicated, treatment of optimal growth models in infinite discrete time horizon, (b) to train the reader to the use of optimal growth models and hence to help them to go further in their research. We are convinced that there is a place for a book which stays somewhere between the "minimum tool kit" and specialized monographs leading to the frontiers of research on optimal growth.

The book being intented for mathematicians as well as for economists, the first chapter provides an introduction to optimal growth models through the "canal" of well-known growth models (Harrod, Domar, Solow-Swan).

Chapter 2 gives a complete treatment of the seminal Ramsey model. This chapter is important to understand the remaining of the book. One of the main results of the Ramsey model is that any optimal path converges to a steady state. The result relies on the assumption that returns are decreasing.

Chapter 3 drops this assumption. It is assumed that the production function convex for low level capital stocks and concave for high level capital stocks. The model, then exhibits a "poverty trap", i.e., a critical value for the initial capital stock. If the initial capital stock is smaller than the critical value, then it will "collapse" (converges to zero), and when it is above that value, it converges to a steady state.

In chapter 4, we provide a framewok for studying standard neoclassical multi-sector models as well endogenous growth models. The

model may be multisectorial, the technology may exhibit any kind of returns and also growth. One of the restrictions of this chapter is the continuity of the instantaneous utility function. It could not take value equals to $-\infty$.
In chapter 5, we allow the utility to be $-\infty$ at some points. We have in mind utility functions used by macroeconomists, as Logarithm or isoelastic functions $(u(c) = \frac{c^\mu}{\mu}, \mu \neq 0)$.
Chapter 6 gives an equivalence between optimal growth paths and competitive equilibrium. It is a formulation, in the frame of intertemporal optimal growth models, of the First and Second Welfare Theorems.
In chapter 7, we study the Ramsey model when there is no discounting, i.e., the discount factor equals 1. Recall that the "true" Ramsey model is without discounting.
Chapter 8 is one the most difficult chapters of the book. We prove that when the discount factor is close to 1, then the optimal capital stocks sequence converges to a steady state different from zero. This poperty is usually called "turnpike".
In chapter 9, we show that complex dynamics (cycles, chaos) could arise in optimal growth models.
A chapter, called Appendix, is devoted for mathematical preliminaries.
We are grateful to the University of Paris I and Ecole Nationale de la Statistique et de l'Administration Economique (ENSAE) for having given us the opportunity to teach part of this course to their PhD students.
We would like to thank Charles-Henri Dimaria, Pascal Gourdel, Lionel Ragot, John Stachurski and Yiannis Vailakis for carefully reading the manuscript and providing numerous corrections and advices.

Paris
November 2002

Cuong Le Van
Rose-Anne Dana

1. Some Growth Models

All the models we consider in this chapter are general equilibrium models. We assume that households own firms and the technics of production. Firms rent the factors of production and produce goods. At each period, all markets clear. We further assume that the techniques of production are identical for the whole economy. The good which is produced at date t may be consumed at the same date or used as capital for the next period. The economy under consideration is a closed economy. In such an economy, production equals national income, and investment equals saving.
In this chapter, we will present a number of growth models to introduce the readers to the problems involved in optimal growth models. There are, essentially four kinds of questions. (i) How the agents determine their consumption and hence the saving which is necessary for investment,(ii) what are the sources of growth, (iii) do the *balanced growth paths* (the definition of which will be given below) exist in these models, and, finally, (iv) stability of the balanced growth path, i.e., when the starting point of the economy is not on the balanced growth path, does this economy converge, in the long term, to the balanced growth path.

We will focus on four models respectively proposed by Harrod [42],[43] , Domar [33],[34], Solow [77] and Swan [82], and Frankel [38]. All these models assume that the *rate of saving is exogenous and constant over time.* In Harrod's model, the technics of production are of Leontief type. Domar's model is, in fact, a particular case of Harrod's model. The investment function in Domar model implicitly assumes that the production function is of Leontief type. In Solow [77], the production function is Cobb-Douglas. Frankel[38] considers a Cobb-Douglas production function and assumes that the total productivity depends on externalities which are represented by the capital stock and the level of labour. The production function which incorporates these externalities could be concave or convex with respect to capital stock, according to the intensity of the externality.

The chapter is organized as follows: in Section 1.1, we present a synthetic frame of these four models. We define the balanced growth path and its stability. We then present, successively, in Sections 1.2, 1.3, 1.4 and 1.5 the four models. A brief comment will be given in Section 1.6.

1.1 A General Growth Model

We consider an intertemporal closed economy where , at date t, the output will be denoted by Y_t, the capital stock by K_t, the saving by S_t, the investment by I_t, the labour by L_t. The equations of the model are

$$S_t = I_t \tag{1.1}$$

$$S_t = sY_t \tag{1.2}$$

$$K_{t+1} = (1 - \delta)K_t + I_t \tag{1.3}$$

K_0, the initial capital stock, being given,

$$L_t = L_0(1 + n)^t \tag{1.4}$$

Equation (1.1) postulates that saving equals investment; in equation (1.2), s is the rate of saving, the value of which is in the interval $[0, 1]$; relation (1.3) generates the dynamics of the sequence of capital stocks; the parameter δ is the rate of depreciation of capital and hence, belongs to $[0, 1]$; relation (1.4) implies that the dynamics of the labour force is exogenous to the model: labour force grows at rate $n > 0$.

To complete the model, we add one of the two following equations:

$$Y_t = F_t(K_t, L_t) \tag{1.5}$$

$$I_t = \Phi(Y_t, Y_{t-1}, ...) \tag{1.6}$$

and Y_0 is given. Equation (1.5) means that the output is produced via a production function F_t using two factors which are capital and labour. Equation (1.6) postulates that the investment depends on the sequence of output $(Y_t, Y_{t-1}, ..)$ which is an indicator of the expected output at date t.

let us introduce a notation which will be used throughout this chapter. An infinite sequence of real numbers $(x_0, x_1, ..., x_t, ...)$ will be denoted by $\mathbf{x}$.

We say that the set of sequences $(\mathbf{Y}, \mathbf{K}, \mathbf{I}, \mathbf{S})$ is a *balanced growth path* of our model if
(i) they are a solution to our model and
(ii) there exist positive numbers (Y^s, K^s, I^s, S^s) and real numbers g_Y, g_K, g_I, g_S such that $\forall t \geq 0$,

$$Y_t = Y^s(1+g_y)^t, K_t = K^s(1+g_k)^t, I_t = I^s(1+g_I)^t, S_t = S^s(1+g_S)^t.$$

In other words there exists a solution which grows at constant rate. The numbers g_Y, g_K, g_I, g_S are respectively the rates of growth of output, capital stock, investment and saving.
The set (Y^s, K^s, I^s, S^s) will be called a *steady state* of the model. Indeed, if the values at date 0 of these variables equal the steady state, then the solution to the model with these initial data will remain on the balanced growth path.
In the models we study in this chapter, the rate of saving s is constant over time. So, we have answered question (i) on the way of determining the consumption of the households. It is obvious from equations (1.1), (1.2) and (1.3) that, if there exists a balanced growth path $(\mathbf{Y}, \mathbf{K}, \mathbf{I}, \mathbf{S})$, then the rates of growth must be equal. Denote by g this common value.

To study the *stability* of the balanced growth path, consider a path $(\mathbf{Y}, \mathbf{K}, \mathbf{I}, \mathbf{S})$ with initial data different from the steady state. We define a set of auxiliary variables $(\hat{\mathbf{Y}}, \hat{\mathbf{K}}, \hat{\mathbf{I}}, \hat{\mathbf{S}})$ by

$$\forall t, \hat{Y}_t = \frac{Y_t}{(1+g)^t}, \hat{K}_t = \frac{K_t}{(1+g)^t}, \hat{I}_t = \frac{I_t}{(1+g)^t}, \hat{S}_t = \frac{S_t}{(1+g)^t}.$$

It is obvious that the balanced growth path is stable if, and only if, for any set of initial data, the sequences $(\hat{\mathbf{Y}}, \hat{\mathbf{K}}, \hat{\mathbf{I}}, \hat{\mathbf{S}})$ converges to the steady state when t converges to $+\infty$.

1.2 The Harrod Model

In this model, the production function is of Leontief type. Equation (1.5) becomes

$$Y_t = \min\{\frac{K_t}{v}, \frac{L_t}{\alpha}\} \tag{1.7}$$

where $v > 0, \alpha > 0$. Let us compute the balanced growth path. Write

$$Y_t = Y^s(1+g)^t, K_t = K^s(1+g)^t, I_t = I^s(1+g)^t, S_t = S^s(1+g)^t.$$

It is easy to check that:

$$g = \frac{s}{v} - \delta \leq n; K^s = vY^s; I^s = S^s = sY^s;$$

There exists a balanced growth path if, and only if, the exogenous parameters s, v, δ, n satisfy the relation $\frac{s}{v} - \delta \leq n$. Under this condition, the rate of growth g is large if either n is, or if s is or if v is small. In particular, if the capital productivity is high, i.e., v is small, then the rate of growth of the economy is high. Since the steady state (Y^s, K^s, I^s, S^s) is determined by three relations, one concludes that there exists an infinity of steady states in the Harrod model.

In fact, one can easily solve the Harrod model. Given the initial value K_0 of the capital stock, we obtain for all $t \geq 0$,

$$Y^t = \frac{K_0}{v}(1+g)^t, K^t = K_0(1+g)^t,$$

$$I^t = K_0\frac{s}{v}(1+g)^t, S^t = K_0\frac{s}{v}(1+g)^t.$$

The solution is therefore proportionnal to the initial value of the capital stock.

1.3 The Domar Model

In this model, the production function is not given. We have an investment function, i.e., equation (1.6). Here it is:

$$I_t = v(Y_{t+1} - (1-\delta)Y_t) \tag{1.8}$$

But observe that, from Leontief production function of the Harrod model (equation (1.7)), we have

$$Y_{t+1} - (1-\delta)Y_t = \frac{K_{t+1} - (1-\delta)K_t}{v} = \frac{I_t}{v}$$

which is exactly the investment function of the Domar model. This investment function is called an *accelerator*. The consumption C_t is equal to $(1-s)Y_t$. If we rewrite equation (1.1) as:

$$Y_t = C_t + I_t,$$

with the forms of the consumption and of the investment functions, we recognize that the Domar model is based on the well-known multiplier-accelerator principle in Macroeconomics. Observe that in the Domar model, labour has no role.

The reader can easily check that steady states are determined, as before, by three relations

$$I^s = S^s, I^s = sY^s, Y^s = \frac{K^s}{v},$$

and the rate of growth of a balanced growth path is $g = \frac{s}{v} - \delta$, the same value as in the Harrod model. Since labor has no role, there is no relationship between the rate of growth of labour force and the rate of growth of the output. One can easily prove that, as in the Harrod model, there exists an infinity of balanced growth paths. Given K_0 and Y_0, tedious computations give the solution of the Domar model:

$$Y_t = (1 + \frac{s}{v} - \delta)^t Y_0,$$

$$K_t = (1-\delta)^t (K_0 - \frac{sY_0}{g+\delta}) + (\frac{sY_0}{g+\delta})(1 + \frac{s}{v} - \delta)^t.$$

We find that $lim_{t\to+\infty}\frac{K_t}{Y_t} = v$. Unsurprisingly, we find that the accelerator parameter v equals the long-term capital-output ratio. If we define $\hat{Y}_t = \frac{Y_t}{(1+g)^t}, \hat{K}_t = \frac{K_t}{(1+g)^t}$, then we find that $\hat{Y}_t = Y_0, \forall t$ and $\hat{K}_t \to vY_0$ when t converges to $+\infty$. Thus, if Y_0 is different from Y_s, then the solution $\mathbf{Y}$ or $\mathbf{K}$ will not converge to the initial balanced growth path. In other words, any balanced growth path of the Domar model is unstable.

Remark 1.3.1.
Usually, in the economic literature, the Harrod model is called the Harrod-Domar model.

1.4 The Solow-Swan Model

In this model, the production function is Cobb-Douglas. Equation (1.5) becomes

$$Y_t = aK_t^\alpha L_t^{(1-\alpha)}(1+\gamma)^t, \alpha \in]0,1[, a > 0 \tag{1.9}$$

Here γ is the rate of growth of the total productivity.
Equation (1.3) becomes:

$$K_{t+1} = (1-\delta)K_t + saK_t^\alpha L_t^{(1-\alpha)}(1+\gamma)^t \tag{1.10}$$

and there exists one initial data which is K_0. The initial values for $\mathbf{Y}, \mathbf{I}, \mathbf{S}$ are $Y_0 = aK_0^\alpha L_0^{1-\alpha}, I_0 = S_0 = sY_0$.
Let us compute a balanced growth path for $\mathbf{K}$. Write $K_t = K^s(1+g)^t, \forall t$. From equation (1.10), we find:

$$(1+g) = (1+n)(1+\gamma)^{\frac{1}{1-\alpha}}$$

and

$$K^s(1+g) = (1-\delta)K^s + sa(K^s)^\alpha L_0^{(1-\alpha)}.$$

The rate of growth of the balanced growth path, g, depends on n, the rate of growth of labour, and on γ, the rate of growth of total productivity. If γ equals 0, then $g = n$ as in the Harrod model.
It is obvious that there exists a unique positive steady state K^s which is

$$K^s = (\frac{sa}{g+\delta})^{\frac{1}{1-\alpha}} L_0.$$

Observe that if g, in the Solow-Swan model, does not depend on the rate of saving, the steady state K^s depends on it. The higher is the rate of saving, the higher is the value of K^s.
Let

$$\Phi(k) = \frac{(1-\delta)k + saL_0^{1-\alpha}k^\alpha}{1+g}.$$

Then k^s is the positive fixed point of Φ, i.e., $K^s = \Phi(K^s)$. We have $\Phi' > 0, \Phi'(0) = +\infty, \Phi'' < 0$. Define the function Ψ by $\Psi(k) = \frac{\Phi(k)}{k}$ if $k > 0$, and $\Psi(0) = +\infty$. This function is continuous. The function Φ being strictly concave, the function Ψ is decreasing.
We will prove that the balanced growth path is stable. Let $K_0 \neq K^s, K_0 > 0$. Let $\mathbf{K}$ be the capital path of the solution to the Solow-Swan model. Define the path $\hat{\mathbf{K}}$ as follows:

$$\hat{K}_t = \frac{K_t}{(1+g)^t}, \forall t \geq 0.$$

Then it satisfies:

$$\forall t \geq 0, \hat{K}_{t+1} = \Phi(\hat{K}_t).$$

From that and since Φ is increasing, if $K_0 < K^s$, then

$$\Phi(\hat{K}_t) < \Phi(K^s), \forall t \geq 1,$$

i.e.,

$$\hat{K}_{t+1} < K^s, \forall t \geq 0.$$

The same argument applies when $K_0 > K^s$. In that case,

$$\hat{K}_{t+1} > K^s, \forall t \geq 0.$$

Now, since Ψ is decreasing, if $K_0 < K^s$, then

$$\Psi(\hat{K}_t) > \Psi(K^s) = 1, \forall t \geq 0,$$

i.e., $\hat{K}_{t+1} > \hat{K}_t, \forall t \geq 0$. In that case, $\hat{\mathbf{K}}$ converges to K^s when t goes to infinity.
If $K_0 > K^s$, by the same argument we obtain that $\hat{K}_{t+1} < \hat{K}_t, \forall t \geq 0$. Again, $\hat{\mathbf{K}}$ converges to K^s when t goes to infinity. We have proved that the balanced growth path is stable.

1.5 The Frankel Model

Romer [68] is usually considered as one of the founders of the theory of endogenous growth. However, Frankel [38] has attempted in a growth models setting to endogeneize the rate of growth through the productivity.
In the Frankel model, the production function has the form

$$F(K, L) = A(K, L)K^{\alpha}L^{1-\alpha}$$

where $A(K, L) = a\frac{K^{\gamma}}{L^{\gamma'}}, \gamma, \gamma' > 0$. Relation (1.5) becomes

$$F(K_t, L_t) = aK_t^{\theta}L_t^{\theta'}$$

with $\theta = \alpha + \gamma, \theta' = 1 - \alpha - \gamma'$. It is obvious that we must impose $\theta' \geq 0$.
One can use the same method as before to find a balance growth path. First, observe that the dynamics of the capital stocks is

$$K_{t+1} = (1 - \delta)K_t + saK_t^{\theta}L_0^{\theta'}(1 + n)^{\theta' t}.$$

If $\theta = 1$, there exists a balanced growth path if, and only if, either $\theta' = 0$ or $n = 0$. In this case, $g + \delta = saL_0^{\theta'}$. But, as in the Harrod model, there will be an infinity of steady states, and the balanced growth paths are not stable.
Now, assume $\theta \neq 1$. Tedious computations give:

$$1 + g = (1 + n)^{\frac{\theta'}{1-\theta}}$$

and, if the steady state $K^s > 0$ exists, it satisfies

$$K^s(1 + g) = (1 - \delta)K^s + sa(K^s)^{\theta}L_0^{\theta'}.$$

Assume that $n > 0, \theta' > 0$. In that case, we have

$$\frac{\partial g}{\partial \gamma} = (1 + n)^{\frac{\theta'}{1-\theta}}(\frac{Log(1 + n)}{(1 - \theta)^2})\theta' > 0.$$

This shows that the capital, as externality, has a positive effect on the rate of growth. Let

$$\Phi(k) = \frac{(1-\delta)k + sak^{\theta}L_0^{\theta'}}{1+g}.$$

Then K^s is a fixed point of Φ, i.e.,

$$K^s = (\frac{sa}{g+\delta})^{\frac{1}{1-\theta}} L_0^{\frac{\theta'}{1-\theta}}.$$

Define the sequence $\hat{\mathbf{K}}$ by $\forall t, \hat{K}_t = \frac{K_t}{(1+g)^t}$. Then we have

$$\forall t \geq 0, \hat{K}_{t+1} = \Phi(\hat{K}_t).$$

If $\theta < 1$, then Φ is concave, and as for the Solow-Swan model, the sequence $\hat{\mathbf{K}}$ converges to K^s when t converges to $+\infty$.
Now, consider the case where $\theta > 1$. In that case, $g \leq 0$. Now, the function Φ is strictly convex. We have $\Phi'(0) = \frac{1-\delta}{1+g} = (1-\delta)(1+n)^{\frac{\theta'}{\theta-1}}$.
If $\Phi'(0) \geq 1$, we have $1-\delta \geq 1+g$, i.e., $g+\delta \leq 0$. The steady state K^s does not exist. By the strict convexity of Φ, we have $\forall k > 0, \Phi(k) > k$. Hence, if $K_0 > 0$, then $\hat{K}_{t+1} > \hat{K}_t, \forall t \geq 0$. If the sequence $\hat{\mathbf{K}}$ converges to some $K^* > 0$ we will have $\Phi(K^*) = K^*$, which is a contradiction. The sequence $\hat{\mathbf{K}}$ must converge to $+\infty$.
If $\Phi'(0) < 1$, then K^s exists and is unique. By the strict convexity of Φ, if $0 < k < K^s$, then $\frac{\Phi(k)}{k} < \frac{\Phi(K^s)}{K^s} = 1$, and if $k > K^s$, then $\frac{\Phi(k)}{k} < \frac{\Phi(K^s)}{K^s} = 1$. Hence, if $K_0 < K^s$, then $\hat{K}_{t+1} < \hat{K}_t, \forall t$, and the sequence $\hat{\mathbf{K}}$ converges to zero, and if $K_0 > K^s$, then $\hat{K}_{t+1} > \hat{K}_t, \forall t$, and the sequence $\hat{\mathbf{K}}$ converges to $+\infty$.
Summing up, the balanced growth path of the Frankel model is stable if $\theta < 1$, and unstable if $\theta \geq 1$.

Remark 1.5.1. Frankel [38] does not study the case where $\theta > 1$.

1.6 Some Conclusions

1. The models we present in this chapter have two weaknesses: the rate of saving is exogenous, and the rate of growth is also exogenous. In optimal growth models of the next chapters, with decreasing

returns technology, we endogeneize the rate of saving but the rate of growth is still exogenous.
2. In the growth models of this chapter, the rate of growth depends either on on the rate of growth of the total productivity or on the capital-output ratio . Therefore, it is immediate that there are at least two ways to endogenize the rate of growth: the first is to introduce education or training as factor improving the productivity of labor (see, e.g., Lucas [50]), the second is to introduce technology or knowledge to improve the capital productivity, (e.g. Romer [68], [69]).
3. When the technology exhibits decreasing returns to scale as in Solow-Swan model or as in Frankel model with $\theta < 1$, the growth path converges to a steady state. But when the model has a production function which is convex (Frankel model with $\theta > 1$), then there exists a "poverty trap", i.e., a critical value, K_c, such that, if the initial capital stock K_0 is less than K_c, then any growth path starting from K_0 will converge to zero, and when K_0 is larger than K_c, any growth path from K_0 will grow without bound. In fact, we can easily show that, if the production function is convex when the capital stock is smaller than some value K_I, and concave beyond this value, and if intersects the first bisector in two positive points K_s and K^s with $K_s < K^s$, then any growth path from $K_0 < K_s$ will converge to zero, and any growth path from $K_0 > K_s$ will converge to K^s. What we want to point out is that the existence of "poverty trap" is due to the increasing returns to scale of the technology.

2. The Ramsey Model

The models presented in Chapter 1 assume that the rate of saving is exogenous. That is one of their limits. Another limit of these models is that the rate of growth is exogenously determined. In this chapter, we will endogeneize the rate of saving of the households. But we do not solve the question of the exogeneity of the rate of growth. This problem will be studied later, in our chapter on endogenous growth models. The model we present here, is a discrete-time horizon version of the well-known Ramsey model [66] which was formalized in continuous-time horizon. This model has been studied in more details by Cass [21] and Koopmans [44]. The basic idea in the Ramsey model is to introduce an infinitely lived consumer who maximizes an intertemporal utility function of her intertemporal sequence of consumptions. At each date, her consumption is constrained by the maximum output produced by a stock of physical capital, and by the necessity of saving for obtaining a physical capital stock for the next period production process. The main results are that, under some conditions, optimal sequences of capital stocks and of consumptions exist, and converge to an optimal steady state. Moreover, the sequence of optimal capital stocks is monotonic.

2.1 The Model

We consider an economy in which there are, at each period t, L_t identical consumers. We denote by c_t the consumption, at period t, of one consumer. We assume that the number of consumers grow at rate n, i.e., $L_t = L_0(1+n)^t$, for every t. In this economy, there is a social planner whose task is to promote the welfare of its population

. So, she wants to maximize the global utility of the consumers :

$$\max L_0 \sum_{t=0}^{\infty}(1/(1+\rho))^t(1+n)^t u(c_t)$$

Here, the function u is called the static utility function or instantaneous utility function and the parameter ρ is the positive time preference rate. A large value of ρ means that the consumers prefer the present to the future. At each date t, consumption c_t is subject to the constraint:

$$L_t c_t + I_t \leq F_t(K_t, L_t),$$

where I_t is the investment, F_t is the production function, K_t is the capital stock , L_t is the number of workers (we implicitly assume that the consumers and the workers are physically identical). The capital stock of period $t+1$ is defined by:

$$K_{t+1} = K_t(1-\delta) + I_t,$$

where $\delta \in]0,1[$ is the depreciation rate of the capital stock. Let us assume that the production function F_t exhibits constant returns to scale and let us introduce the per capita capital stock $k_t = K_t/L_t$. The constraint for each period, between consumption and investment becomes:

$$c_t + k_{t+1}(1+n) - (1-\delta)k_t \leq F_t(k_t, 1).$$

Assume that $F_t(k_t, 1) = A(1+\gamma)^t k_t^\alpha$, with $0 < \alpha < 1$. The parameter γ is the rate of growth of the productivity. We then obtain:

$$c_t + k_{t+1}(1+n) \leq A(1+\gamma)^t k_t^\alpha + (1-\delta)k_t.$$

If the utility function u is strictly increasing, then, at the optimum, the constraints will be binding at each period. If the optimal sequences of capital stock and consumption grow at rate g, i.e., for any t, $k_t = k_0(1+g)^t$, $c_t = c_0(1+g)^t$, we then have

$$(1+g)^{(1-\alpha)} = 1+\gamma.$$

In other words, the rate of growth of the economy is determined by the exogenous rate of growth of the productivity. Using the variables capital per capita k_t and consumption per capita c_t, the Ramsey model can be written as:

$$\max \sum_{t=0}^{\infty} \beta^t u(c_t)$$

under the constraints:

$$\forall t, c_t + k_{t+1}(1+n) \leq A k_t^{\alpha} + (1-\delta)k_t,$$

and k_0 is given, and by definition, $\beta = (1+n)/(1+\rho)$. The parameter β will be called discount factor. If we assume, for simplicity, that $n = 0$, and if we define the function f by $f(k) = Ak^{\alpha} + (1-\delta)k$, then the Ramsey model will have the following compact form:

$$\max \sum_{t=0}^{\infty} \beta^t u(c_t)$$

under the constraints:

$$\forall t, c_t + k_{t+1} \leq f(k_t),$$

$$\forall t, c_t \geq 0, k_t \geq 0,$$

and $k_0 \geq 0$ is given. In this chapter, we will make use of this form.

2.1.1 The Assumptions

The following assumptions will be maintained throughout this chapter.

H0 $0 < \beta < 1$.
H1 The function $u : \mathbb{R}_+ \to \mathbb{R}_+$, is twice continuously differentiable and satisfies $u(0) = 0$. Moreover, its derivatives satisfy $u' > 0$ (strictly increasing) and $u'' < 0$ (strictly concave).
H2 Inada Condition : $u'(0) = +\infty$.
H3 The function $f : \mathbb{R}_+ \to \mathbb{R}_+$ is twice continuously differentiable and satisfies $f(0) = 0$. Its derivatives satisfy $f' > 0$ (strictly increasing), $f'' < 0$ (strictly concave), $\lim_{x \to +\infty} f'(x) < 1, f'(0) = M \leq +\infty$.

2.1.2 Feasible Paths

A sequence $(c_t, k_t)_{t=0,\ldots,+\infty}$ is feasible from k_0 if it satisfies:

$$\forall t, c_t + k_{t+1} \leq f(k_t),$$
$$c_t \geq 0, \ k_t \geq 0,$$

and k_0 is given.

A sequence of consumptions $(c_t)_{t=0,\ldots,+\infty}$ is feasible from k_0 if there exists a sequence of capital stocks $(k_t)_{t=0,\ldots,+\infty}$ such that, together, they satisfy the previous relations.

Since the utility function is strictly increasing, at the optimum, the constraints will be binding. Thus, the initial optimal growth model is equivalent to the following problem:

$$\max \sum_{t=0}^{+\infty} \beta^t u(f(k_t) - k_{t+1})$$

under the constraints:

$$\forall t, \ 0 \leq k_{t+1} \leq f(k_t),$$

and $k_0 \geq 0$ is given.

We say that the sequence of capital stocks $(k_t)_{t=0,\ldots,+\infty}$ is feasible from $k_0 \geq 0$ if we have

$$\forall t, \ 0 \leq k_{t+1} \leq f(k_t).$$

We will denote by $\mathbf{k}$ a feasible sequence of capital stocks from k_0 and by $\Pi(k_0)$ the sets of feasible capital $\mathbf{k}$. It may be easily verified that if $k_0 < k_0'$, then $\Pi(k_0) \subset \Pi(k_0')$.

2.2 Existence of Optimal Paths

To prove the existence of optimal capital and optimal consumption paths, we show that the initial optimization problem turns out to be a maximization problem of a continuous function on a compact set. We use, for this purpose, the product topology (for a definition, see the Appendix) defined on the space of infinite sequences of real numbers.

The main result of this section is:

Proposition 2.2.1. *There exists an optimal path* $\mathbf{k}$. *The associated optimal consumption path* $\mathbf{c}$ *is*

$$c_t = f(k_t) - k_{t+1}, \forall t.$$

Proof. The proof will be done in several steps.
Step 1. If $f'(0) \leq 1$, let $\bar{k} = 0$. If $f'(0) > 1$, let $\bar{k}$ be such that $f(\bar{k}) = \bar{k}$ and $\bar{k} > 0$. Such a point exists since $f'(+\infty) < 1$. Observe that $k > \bar{k}$ implies that $f(k) < k$.
If $k_0 \leq \bar{k}$, then we have for $\mathbf{k} \in \Pi(k_0)$:

$$0 \leq k_1 \leq f(k_0) \leq f(\bar{k}) = \bar{k},$$

$$0 \leq k_2 \leq f(k_1) \leq f(\bar{k}) = \bar{k}...$$

That means that $\forall t, 0 \leq k_t \leq \bar{k} = \max(k_0, \bar{k})$.
If $k_0 > \bar{k}$, then $k_1 \leq f(k_0) < k_0$. If $k_1 > \bar{k}$, then $k_2 \leq f(k_1) < k_1 < k_0$. If $k_1 \leq \bar{k}$, then $k_2 \leq f(k_1) \leq \bar{k}$, and hence for $t \geq 2$, $k_t \leq \bar{k}$. So, we have proved that, if $k_0 > \bar{k}$, then $k_t \leq \max(k_0, \bar{k}) = k_0, \forall t$. Thus, for any $\mathbf{k} \in \Pi(k_0)$, we have $0 \leq k_t \leq \max(k_0, \bar{k}), \forall t$. Define $A = \max(k_0, \bar{k})$. Thus, if $\mathbf{k} \in \Pi(k_0)$, then $(k_t)_{t=0,..,+\infty} \in [0, A]^\infty$. The set $[0, A]^\infty$ is compact for the product topology defined on the space of sequences $\mathbf{k}$. Hence, the set of feasible sequences $\Pi(k_0)$ is included in a compact set for the product topology.
Step 2. We will prove that the set of feasible sequences $\Pi(k_0)$ is closed for the product topology, and therefore, is compact for this topology.
Let $(k_0, k_1^n, ..., k_t^n, ...)$ be a feasible sequence which converges to $(k_0, k_1, ..., k_t, ..)$ when n converges to infinity. By the very definition of the product topology, we have that, for every t, k_t^n converges to k_t, when n converges to infinity. Since, for every t:

$$\forall n, 0 \leq k_{t+1}^n \leq f(k_t^n),$$

we have: $0 \leq k_{t+1} \leq f(k_t)$. Thus, we have proved the closeness of the set of feasible sequences $\Pi(k_0)$.
Step 3. We introduce the function U defined on the set of feasible sequences:

$$U(\mathbf{k}) = \sum_{t=0}^{\infty} \beta^t u(f(k_t) - k_{t+1}).$$

The function U is well-defined since $0 \leq U(\mathbf{k}) \leq u(f(A))/(1-\beta)$. We will show that it is continuous for the product topology.
Let $\mathbf{k^n}$ converge to $\mathbf{k}$. We have

$$|U(\mathbf{k^n}) - U(\mathbf{k})| \leq \sum_{t=0}^{+\infty} \beta^t |u(f(k_t) - k_{t+1}) - u(f(k_t^n) - k_{t+1}^n)|.$$

Let $\varepsilon > 0$ be given. There exists T such that for every n

$$0 \leq \sum_{t=T+1}^{+\infty} \beta^t u(f(k_t^n) - k_{t+1}^n) \leq \varepsilon/3,$$

and

$$0 \leq \sum_{t=T+1}^{\infty} \beta^t u(f(k_t) - k_{t+1}) \leq \varepsilon/3,$$

since we have: $\forall n, \forall t, 0 \leq k_t \leq A$, $0 \leq k_t^n \leq A$.
Then

$$|U(\mathbf{k^n}) - U(\mathbf{k})| \leq \sum_{t=0}^{t=T} \beta^t |u(f(k_t) - k_{t+1}) - u(f(k_t^n) - k_{t+1}^n)| + (2/3)\varepsilon.$$

Since the functions u and f are continuous, there exists an integer N such that, for every $n \geq N$, the first term of the second member of the previous inequality is less than $\varepsilon/3$. We have proved that U is continuous.
Step 4. It is obvious that the optimal growth problem is now equivalent to

$$\max\{U(\mathbf{k}) : \mathbf{k} \in \Pi(k_0)\}.$$

Since, for the product topology, the function U is continuous and the set $\Pi(k_0)$ is compact, the optimal growth model has a solution. Since the utility function u is strictly increasing, the optimal consumptions path $\mathbf{c}$ must satisfy

$$\forall t, c_t = f(k_t) - k_{t+1}.$$

□

2.3 Properties of the Optimal Paths

In this section, we will prove: first, the uniqueness of the solution $(\mathbf{k}, \mathbf{c})$ to the optimal growth model, and second, that it satisfies $\forall t, c_t > 0,\ k_t > 0$.

Uniqueness of optimal paths follows from the strict concavity of the utility function u, while the positiveness of optimal consumptions and capital stocks follows from the Inada condition ($u'(0) = +\infty$). Using this last property, we show that the optimal capital stocks satisfy an important relation called the **Euler equation**.

Proposition 2.3.1. *(i) Given k_0, there exists a unique solution $(\mathbf{k}, \mathbf{c})$ to the optimal growth model.*
(ii) This solution satisfies $\forall t, c_t > 0,\ k_t > 0$, if $k_0 > 0$.
(iii) When $k_0 > 0$, the solution satisfies the Euler equation

$$\forall t, u'(f(k_t) - k_{t+1}) = \beta u'(f(k_{t+1}) - k_{t+2}) f'(k_{t+1}). \qquad (\mathcal{E})$$

Proof. (i) Assume that there exist two different optimal capital paths $\mathbf{k^1}$, $\mathbf{k^2}$. Let $\mathbf{c^1}$, $\mathbf{c^2}$ be the associated consumption paths, i.e.,

$$\forall t, c_t^1 + k_{t+1}^1 = f(k_t^1),$$

$$\forall t, c_t^2 + k_{t+1}^2 = f(k_t^2).$$

There must be some t such that $c_t^1 \neq c_t^2$. If not, we have $\mathbf{k^1} = \mathbf{k^2}$. Let $\lambda \in]0,1[$. Then we have

$$\forall t, \lambda c_t^1 + (1-\lambda) c_t^2 + \lambda k_{t+1}^1 + (1-\lambda) k_{t+1}^2 \leq f(\lambda k_t^1 + (1-\lambda) k_t^2).$$

The sequence $(\lambda \mathbf{k^1} + (\mathbf{1} - \lambda)\mathbf{k^2})$ belongs to $\Pi(k_0)$. We have

$$\sum_{t=0}^{+\infty} \beta^t u(f(\lambda k_t^1 + (1-\lambda) k_t^2) - \lambda k_{t+1}^1 - (1-\lambda) k_{t+1}^2)$$

$$\geq \sum_{t=0}^{+\infty} \beta^t u(\lambda c_t^1 + (1-\lambda) c_t^2)$$

$$> \sum_{t=0}^{+\infty} \beta^t(\lambda u(c_t^1) + (1-\lambda)u(c_t^2)) = \sum_{t=0}^{+\infty} \beta^t u(c_t^1),$$

since there exists t such that $c_t^1 \neq c_t^2$, and hence, $u(\lambda c_t^1 + (1-\lambda)c_t^2) > \lambda u(c_t^1) + (1-\lambda)u(c_t^2)$ by the strict concavity of u. We obtain a contradiction.

(ii) Since $k_0 > 0$, the feasible consumption path $(0, 0, ..., 0, ...)$ is not optimal since u is strictly increasing. Hence, there exists some t with $c_t > 0$. For simplicity, assume $c_0 = 0$ and $c_1 > 0$. We have $k_1 = f(k_0)$ and $c_1 + k_2 = f(k_1)$. Since $k_0 > 0$, we have $k_1 > 0$. Choose some $\varepsilon > 0$ such that $k_1 - \varepsilon > 0$ and $f(k_1 - \varepsilon) - k_2 > 0$. Define sequences $(\mathbf{c}', \mathbf{k}')$ as follows:

$$c_0' = \varepsilon, c_1' = f(k_1 - \varepsilon) - k_2,$$

and

$$\forall t \geq 2, \; c_t' = c_t, k_t' = k_t.$$

These sequences are feasible from k_0. Compute

$$\Delta_\varepsilon = \sum_{t=0}^{+\infty} \beta^t u(c_t') - \sum_{t=0}^{+\infty} \beta^t u(c_t).$$

We find:

$$\begin{aligned} \Delta_\varepsilon &= u(c_0') + \beta u(c_1') - u(c_0) - \beta u(c_1) \\ &\geq u'(c_0')\varepsilon + \beta u'(c_1')(f(k_1 - \varepsilon) - f(k_1)) \\ &= \varepsilon[u'(c_0') + \beta u'(c_1')(f(k_1 - \varepsilon) - f(k_1))/\varepsilon]. \end{aligned}$$

By the Inada Condition, when $\varepsilon \to 0$, $u'(c_0') \to +\infty$, while $u'(c_1')(f(k_1 - \varepsilon) - f(k_1))/\varepsilon \to -u'(c_1)f'(k_1) > -\infty$. Hence, $\Delta_\varepsilon > 0$ when ε is small enough. We obtain a contradiction. We have proved that $c_0 > 0$.

We now claim that $k_1 > 0$. Assume the contrary: $k_1 = 0$. In this case $k_t = 0$, $c_t = 0$, $\forall t \geq 1$. As previously, define the sequences $(\mathbf{c}', \mathbf{k}')$ as follows :

$$c_0' = c_0 - \varepsilon, k_1' = \varepsilon, c_1' = f(\varepsilon),$$

and

$$\forall t \geq 2, c'_t = c_t, k'_t = k_t,$$

where ε is chosen such that $\varepsilon > 0$ and $c'_0 > 0$. Let us compute

$$\Delta_\varepsilon = \sum_{t=0}^{+\infty} \beta^t u(c'_t) - \sum_{t=0}^{+\infty} \beta^t u(c_t).$$

We find

$$\Delta_\varepsilon = u(c'_0) + \beta u(c'_1) - u(c_0) - \beta u(c_1)$$

$$\geq u'(c_0 - \varepsilon)(-\varepsilon) + \beta u'(f(\varepsilon))f(\varepsilon) = \varepsilon[\beta u'(f(\varepsilon))f(\varepsilon)/\varepsilon - u'(c_0 - \varepsilon)].$$

When $\varepsilon \to 0$, we have $u'(f(\varepsilon))f(\varepsilon)/\varepsilon \to +\infty$, and $u'(c_0 - \varepsilon) \to u'(c_0) < +\infty$. Hence, $\Delta_\varepsilon > 0$ when ε is small enough. Again, we have a contradiction. To sum up, we have proved that if $c_0 > 0$ then $k_1 > 0$. But, following the proof of the positiveness of c_0, we must have $c_1 > 0$ since $k_1 > 0$. By induction, $c_t > 0$, $k_t > 0$, for every t.
(iii) From (ii), we have $\forall t \geq 0, 0 < k_{t+1} < f(k_t)$. Consider the following sequence $\mathbf{k}'$ defined as follows:

$$\forall \tau \neq t+1, k'_\tau = k_t; k'_{t+1} = y,$$

where y is in open neighborhood of k_{t+1} and satisfies $0 < y < f(k_t), 0 < k_{t+2} < f(y)$. The sequence $\mathbf{k}'$ is feasible from k_0. Hence $U(\mathbf{k}) \geq U(\mathbf{k}')$. This inequality is equivalent to

$$u(f(k_t)-k_{t+1})+\beta u(f(k_{t+1})-k_{t+2}) \geq u(f(k_t)-y)+\beta u(f(y)-k_{t+2}).$$

It holds for any y in this neighborhood. Thus, the function $\Phi(y) = u(f(k_t) - y) + \beta u(f(y) - k_{t+2})$ has a local maximum at k_{t+1}. By writing $\Phi'(k_{t+1}) = 0$, we then obtain the Euler equation.

2.4 Value Function - Bellman Equation - Optimal Policy

We have shown that the Ramsey model is equivalent to the maximization of an objective function U defined on the set of feasible sequences from the initial capital stock k_0. The maximal value of this problem depends on the initial value k_0 and is called *Value function* of k_0. More precisely, we define the Value function V by:

$$\forall k_0 \geq 0, V(k_0) = \max \sum_{t=0}^{+\infty} \beta^t u(f(k_t) - k_{t+1})$$

under the constraints:

$$\forall t,\ 0 \leq k_{t+1} \leq f(k_t),$$

and $k_0 \geq 0$ is given.

2.4.1 Some Properties of the Value Function

We will show that the Value function is strictly concave, continuous and strictly increasing.

Proposition 2.4.1. *The Value function satisfies:*
(i) $V(0) = 0; V(k_0) > 0$, *if* $k_0 > 0$. *If* $\mathbf{k}$ *is the optimal path, then* $V(k_0) = \sum_{t=0}^{+\infty} \beta^t u(f(k_t) - k_{t+1})$.
(ii) V *is strictly concave, continuous.*
(iii) V *is strictly increasing.*
(iv) Let $k_0 > 0$ *and* $\mathbf{k}$ *be the optimal path from* k_0. *The Value function* V *is differentiable at* k_0. *We have* $V'(k_0) = u'(f(k_0) - k_1)f'(k_0)$.

Proof. (i) The proof of this statement is straightforward.
(ii) Let $\mathbf{k}$ be feasible from k_0, and $\mathbf{k}'$ be feasible from k'_0, with $k'_0 \neq k_0$. Let $\lambda \in]0,1[$. We have for any $t \geq 0$:

$$0 \leq \lambda k_{t+1} + (1-\lambda)k'_{t+1} \leq \lambda f(k_t) + (1-\lambda)f(k'_t) \leq f(\lambda k_t + (1-\lambda)k'_t).$$

Hence,

$$V(\lambda k_0 + (1-\lambda)k'_0) \geq \sum_{t=0}^{+\infty} \beta^t u(f(\lambda k_t + (1-\lambda)k'_t) - \lambda k_{t+1} - (1-\lambda)k'_{t+1})$$

$$\geq \sum_{t=0}^{+\infty} \beta^t u[\lambda(f(k_t) - k_{t+1}) + (1-\lambda)(f(k'_t) - k'_{t+1})]$$

$$\geq \lambda \sum_{t=0}^{+\infty} \beta^t u(f(k_t) - k_{t+1}) + (1-\lambda) \sum_{t=0}^{+\infty} \beta^t u(f(k'_t) - k'_{t+1})$$

$$= \lambda V(k_0) + (1-\lambda) V(k'_0).$$

If $k'_0 = 0$, then $k_0 > 0$. From Proposition 2.3.1 $f(k_t) - k_{t+1} > 0,\ \forall t$. We also have $\mathbf{k}' = 0$. Hence,

$$\forall t, u(\lambda(f(k_t) - k_{t+1}) + (1-\lambda)(f(k'_t) - k'_{t+1})) >$$

$$\lambda u(f(k_t) - k_{t+1}) + (1-\lambda) u(f(k'_t) - k'_{t+1}),$$

and this implies $V(\lambda k_0 + (1-\lambda)k'_0) > \lambda V(k_0) + (1-\lambda)V(k'_0)$.
Now assume $k_0 > 0,\ k'_0 > 0$. We will show that there exists t with $f(k_t) - k_{t+1} \neq f(k'_t) - k'_{t+1}$. Indeed, on the contrary, we have $f(k_0) - k_1 = f(k'_0) - k'_1$ and $f(k_1) - k_2 = f(k'_1) - k'_2$. By Euler equation (see Proposition 2.3.1) we have $f'(k_1) = f'(k'_1)$, i.e., $k_1 = k'_1$,and therefore, $k_0 = k'_0$, in contradiction with the initial assumption. From the strict concavity of u, for such a t, we have

$$u(\lambda(f(k_t) - k_{t+1}) + (1-\lambda)(f(k'_t) - k'_{t+1})) >$$

$$\lambda u(f(k_t) - k_{t+1}) + (1-\lambda) u(f(k'_t) - k'_{t+1}),$$

and then $V(\lambda k_0 + (1-\lambda)k'_0) > \lambda V(k_0) + (1-\lambda)V(k'_0)$. We have proved the strict concavity of V.
Since V is concave on $[0, +\infty[$, it is continuous in $]0, +\infty[$ (see the Appendix). We just have to show that V is continuous at the origin. Let k_0 converge to zero. When k_0 is small enough we have $k_0 \leq \bar{k}$, where $\bar{k}$ satisfies $f(\bar{k}) = \bar{k}$. If $\mathbf{k}$ is feasible from k_0, then:

$$\forall t,\ 0 \leq k_t \leq f^t(k_0) \leq f^t(\bar{k}) = \bar{k}.$$

For any T, the following inequality holds:

$$\sum_{T+1}^{+\infty} \beta^t u(f(k_t) - k_{t+1}) \leq \sum_{T+1}^{+\infty} \beta^t u(f(k_t)) \leq u(\bar{k})(\beta^{T+1}/(1-\beta)).$$

Let $\varepsilon > 0$ be given. For T large enough, we have

$$\sum_{T+1}^{+\infty} \beta^t u(f(k_t) - k_{t+1}) \leq (\varepsilon/2).$$

Fix such a T. If k_0 is small enough, we get

$$\sum_{t=0}^{T} \beta^t u(f(k_t) - k_{t+1}) \leq \sum_{t=0}^{T} \beta^t u(f^t(k_0)) \leq (\varepsilon/2).$$

To sum up, for k_0 small enough, for any $\mathbf{k}$ feasible from k_0, one has $\sum_{t=0}^{+\infty} \beta^t u(f(k_t) - k_{t+1}) \leq \varepsilon$, and hence, $V(k_0) \leq \varepsilon$. This ends the proof of the continuity of V.

(iii) Let $k_0 < k_0'$ and $\mathbf{k}$ be the optimal path from k_0. We have $V(k_0) = \sum_{t=0}^{+\infty} \beta^t u(f(k_t) - k_{t+1})$. The sequence $\mathbf{k}$ satisfies

$$0 < k_1 < f(k_0),$$

$$\forall t \geq 2, \ 0 < k_t < f(k_{t-1}).$$

But $k_1 < f(k_0) < f(k_0')$. Hence

$$V(k_0) < u(f(k_0') - k_1) + \sum_{t=1}^{+\infty} \beta^t u(f(k_t) - k_{t+1}) \leq V(k_0'),$$

since the sequence $(k_0', k_1, ..., k_t, ...)$ is feasible from k_0'.

(iv) Let $k_0 > 0$ be given. The function V is subdifferentiable at k_0 (see the Appendix). As V is concave, to prove that it is differentiable at k_0, it suffices to prove that its subdifferential at k_0 is a singleton. Let $\mathbf{k}$ be the optimal path from k_0. We know that $V(k_0) = \sum_{t=0}^{+\infty} \beta^t u(f(k_t) - k_{t+1})$, and $0 < k_1 < f(k_0)$. If k_0' is in a small neighborhood of k_0, then $k_1 < f(k_0')$ and the sequence $(k_0', k_1, ..., k_t, ...)$ is feasible from k_0'. In this case,

$$V(k_0') \geq u(f(k_0') - k_1) + \sum_{t=1}^{+\infty} \beta^t u(f(k_t) - k_{t+1}).$$

Let p be a subgradient of V at k_0. We have

$$\sum_{t=0}^{+\infty} \beta^t u(f(k_t) - k_{t+1}) - u(f(k'_0) - k_1) + \sum_{t=1}^{+\infty} \beta^t u(f(k_t) - k_{t+1})$$

$$\geq V(k_0) - V(k'_0) \geq p(k_0 - k'_0).$$

But these inequalities give

$$u(f(k_0) - k_1) - u(f(k'_0) - k_1) \geq p(k_0 - k'_0).$$

This shows that p is a subgradient at k_0 of the function $k \to u(f(k) - k_1)$ where k is in the neighborhood of k_0. Since the function $u(f(k) - k_1)$ is differentiable at k_0, we have $p = u'(f(k_0) - k_1)f'(k_0)$ (see the Appendix). This result shows that the subdifferential of V at k_0 is a singleton. Therefore, V is differentiable at k_0 and $V'(k_0) = u'(f(k_0) - k_1)f'(k_0)$. □

2.4.2 Bellman Equation

Proposition 2.4.2. *(i) The Value function V satisfies the Bellman equation:*

$$\forall k_0 \geq 0, V(k_0) = \max\{u(f(k_0) - k) + \beta V(k) : 0 \leq k \leq f(k_0)\} \quad (\mathcal{B})$$

(ii) The Value function V is the unique solution to the Bellman equation which is both continuous and satisfies the transversality condition:

$$\forall k_0 \geq 0, \ \forall \mathbf{k} \in \Pi(k_0), \ \lim_{T \to +\infty} \beta^T V(k_T) = 0 \quad (\mathcal{T})$$

(iii) A sequence $\mathbf{k} \in \Pi(k_0)$ is the optimal solution of the Ramsey model if, and only if, it satisfies:

$$\forall t, \ V(k_t) = u(f(k_t) - k_{t+1}) + \beta V(k_{t+1}).$$

Proof. (i) Let $\mathbf{k}$ be the optimal path from k_0. We have

$$V(k_0) = u(f(k_0) - k_1) + \beta \sum_{t=0}^{+\infty} \beta^t u(f(k_{t+1}) - k_{t+2})$$

$$\leq u(f(k_0) - k_1) + \beta V(k_1),$$

since $(k_1, k_2, ..., k_t, ...)$ is feasible from k_1. Hence

$$V(k_0) \leq \max\{u(f(k_0) - k) + \beta V(k) : 0 \leq k \leq f(k_0)\}.$$

We will prove the converse. Since u and V are continuous, there exists $k_1 \in [0, f(k_0)]$ such that

$$\max\{u(f(k_0)-k)+\beta V(k) : k \in [0, f(k_0)]\} = u(f(k_0)-k_1)+\beta V(k_1).$$

But we know that there exists a sequence $(k'_0, k'_1, ..., k'_t, ...)$ feasible from k_1 (and hence, $k'_0 = k_1$) such that $V(k_1) = \sum_{t=0}^{+\infty} \beta^t u(f(k'_t) - k'_{t+1})$. Consider the following sequence $\mathbf{k}$: $\forall t \geq 1,\ k_{t+1} = k'_t$. This sequence belongs to $\Pi(k_0)$. Therefore we have:

$$u(f(k_0) - k_1) + \beta V(k_1) = u(f(k_0) - k_1) + \beta \sum_{t=0}^{+\infty} \beta^t u(f(k'_t) - k'_{t+1})$$

$$= \sum_{t=0}^{+\infty} \beta^t u(f(k_t) - k_{t+1}) \leq V(k_0).$$

This ends the proof of the statement.

(ii) We first prove that the Value function V satisfies the Transversality Condition. Let $\mathbf{k}$ be in $\Pi(k_0)$. In this case $\forall t,\ 0 \leq k_t \leq \max\{k_0, \bar{k}\}$ (see Step 1 of the proof of Proposition 2.1.1). Similarly, for any sequence $(x_0, x_1, ..., x_t, ...)$ feasible from k_T, we have

$$\forall t,\ 0 \leq x_t \leq \max\{k_T, \bar{k}\} \leq A = \max\{k_0, \bar{k}\}.$$

Since V is increasing, we have $V(k_t) \leq V(A), \forall t$. Hence,

$$\lim_{T \to +\infty} \beta^T V(k_T) = 0.$$

Let W be another solution to the Bellman equation, which is continuous and which satisfies the transversality condition. There exists $k_1 \in [0, f(k_0)]$ such that $W(k_0) = u(f(k_0) - k_1) + \beta W(k_1)$, and by induction, there exists a sequence $(k_1, k_2, ..., k_T, ...)$ which satisfies:

$$0 \leq k_{t+1} \leq f(k_t), \ \forall t \geq 1,$$

and $\forall T$,

$$W(k_0) = u(f(k_0) - k_1) + \beta u(f(k_1) - k_2) + \dots$$

$$+\beta^{T-1} u(f(k_{T-1}) - k_T) + \beta^T W(k_T).$$

Since $\beta^T W(k_T) \to 0$, we have $W(k_0) = \sum_{t=0}^{+\infty} \beta^t u(f(k_t) - k_{t+1})$. Hence $W(k_0) \leq V(k_0)$.

To prove the converse, take a feasible sequence $\mathbf{k}$ from k_0. We have

$$W(k_0) \geq u(f(k_0) - k_1) + \beta W(k_1),$$

$$W(k_1) \geq u(f(k_1) - k_2) + \beta W(k_2),$$

and $\forall T$,

$$W(k_0) \geq u(f(k_0) - k_1) + \beta u(f(k_1) - k_2) + \dots$$

$$+\beta^{T-1} u(f(k_{T-1}) - k_T) + \beta^T W(k_T).$$

Since $\lim_{T \to +\infty} \beta^T W(k_T) = 0$, we have $W(k_0) \geq \sum_{t=0}^{+\infty} \beta^t u(f(k_t) - k_{t+1})$. This inequality holds for any feasible sequence. Therefore, $W(k_0) \geq V(k_0)$.

(iii) Let $\mathbf{k}$ be the optimal sequence of capital stocks. We claim that $V(k_0) = u(f(k_0) - k_1) + \beta V(k_1)$. Since we have $V(k_0) = \sum_{t=0}^{+\infty} u(f(k_t) - k_{t+1})$, and since the sequence $(k_1, k_2, ..., k_t, ...)$ is feasible from k_1, it is obvious that $V(k_0) \leq u(f(k_0) - k_1) + \beta V(k_1)$. But there exists an optimal path from k_1, say $(k_1, x_2, x_3, ..., x_t, ...)$. We have:

$$u(f(k_0) - k_1) + \beta V(k_1)$$

$$= u(f(k_0) - k_1) + \beta u(f(k_1) - x_2) + \sum_{t=2}^{+\infty} \beta^t u(f(x_t) - x_{t+1})$$

$$\leq V(k_0),$$

since the sequence $(k_0, k_1, x_2, ..., x_t, ...)$ is feasible from k_0. We have proved the claim. By induction, $\forall t, V(k_t) = u(f(k_t) - k_{t+1}) + \beta V(k_{t+1})$.

Conversely, let $\mathbf{k}$ be a feasible sequence from k_0 which satisfies

$$\forall t, V(k_t) = u(f(k_t) - k_{t+1}) + \beta V(k_{t+1}).$$

By induction, one gets:

$$V(k_0) = \sum_{t=0}^{T} \beta^t u(f(k_t) - k_{t+1}) + \beta^{t+1} V(k_{t+1}).$$

Hence,$V(k_0) = \sum_{t=0}^{+\infty} \beta^t u(f(k_t) - k_{t+1})$, since $\lim_{t \to +\infty} \beta^t V(k_t) = 0$. This proves the optimality of the sequence $\mathbf{k}$. □

Remark 2.4.1. Let us show, by a more general method, that V is the unique solution to the Bellman Equation. Actually, we show that V is the fixed point of a contraction mapping in the Banach space of bounded functions (see the the Appendix for a definition and for the property of contraction mappings in Banach spaces).

For this purpose, let $\hat{A} > \bar{k}$ and $X = [0, \hat{A}]$. It is obvious that $\forall k_0 \in X, \forall \mathbf{k} \in \Pi(k_0)$, we have $f(k_0) \leq \hat{A}$, and $\forall t,\ k_t \in X$. We restrict V to X. We denote by $B(X)$ the space of bounded functions on X. This space will be endowed with the sup-norm $\|h\| = \sup_{x \in X} |h(x)|$, where $h \in B(X)$.
We denote by $\mathbf{T}$ the operator from $B(X)$ into $B(X)$ defined as follows:

$$\forall h \in B(X), \forall k_0 \in X,$$

$$\mathbf{T}h(k_0) = \sup\{u(f(k_0) - y) + \beta h(y) : y \in [0, f(k_0)]\}.$$

Let us check that $\mathbf{T}$ maps $B(X)$ into $B(X)$.
Indeed, we have

$$|\mathbf{T}h(k_0)| \leq u(f(k_0)) + \beta\|h\| \leq u(\hat{A}) + \|h\|.$$

We claim that $\mathbf{T}$ is a β-contraction mapping. We use the following theorem

Blackwell Theorem
Let T be an operator from $B(X)$ into $B(X)$. Assume that T satisfies:
(i) $h \leq g \Rightarrow Th \leq Tg$,
(ii) There exists $\alpha \in]0, 1[$ such that $T(h + a) \leq Th + \alpha a$, where $a \in \mathbb{R}_+$.
Then T is a α-contraction mapping.

Proof. We have: $\forall x \in X, h(x) \leq g(x) + \|h - g\|$. Hence, $Th(x) \leq Tg(x) + \alpha\|h - g\|, \forall x \in X$. But, we also have $Tg(x) \leq Th(x) + \alpha\|h - g\|, \forall x \in X$. The result follows. □

Obviously, the operator $\mathbf{T}$ satisfies the conditions of Blackwell Theorem. Hence, there exists a unique fixed point in $B(X)$ which is V. One of the well-known properties of contraction mappings is that the fixed-point is the limit of the sequence $(\mathbf{T}^n h)$ when n converges to infinity, where h is any element of $B(X)$. More precisely, we have $\lim_{n \to +\infty} \|\mathbf{T}^n h - V\| = 0$. In particular, we have: $\forall x \in X, V(x) = \lim_{n \to +\infty} T^n h(x)$.
The function V is also the unique solution to the Bellman equation, which is continuous. Indeed, if h is continuous, then it is also bounded on X. From the Maximum Theorem (see the Appendix), $\mathbf{T}h$ is continuous. If h is a solution to the Bellman equation, it must be the fixed point of $\mathbf{T}$ and hence, equals V.
The function V is also the unique solution to the Bellman equation, which is continuous and concave. The argument is the same. Since u and f are concave, if h is concave then $\mathbf{T}h$ is concave. Thus, if h is a fixed point of $\mathbf{T}$, then $h = V$.

2.4.3 Optimal Policy

Let $k_1 = \text{argmax } \{u(f(k_0) - y) + \beta V(y) : y \in [0, f(k_0)]\}$, i.e., k_1 satisfies $k_1 \in [0, f(k_0)]$ and $V(k_0) = u(f(k_0) - k_1) + \beta V(k_1)$. Such a k_1 is unique, since the function $y \rightarrow u(f(k_0) - y) + \beta V(y)$ is strictly concave.
So, we can define a mapping φ which associates with any k_0 the maximizer k_1. We will show that φ is increasing. Moreover, the function $f - \varphi$ is also increasing. In other words, the optimal capital stock and the optimal consumption from k_0 increase with k_0.

Proposition 2.4.3. *(i) The mapping φ is continuous.*
(ii) If $k_0 > 0$ then $\varphi(k_0) > 0$, and $f(k_0) - \varphi(k_0) > 0$.
(iii) A feasible sequence $\mathbf{k} \in \Pi(k_0)$ is optimal if, and only if, it satisfies: $\forall t, k_t = \varphi^t(k_0)$, where $\varphi^2(k_0) = \varphi(\varphi(k_0))$, and by induction, $\varphi^t(k_0) = \varphi(\varphi^{t-1}(k_0))$.
(iv) The function φ is increasing.
(v) The function $f - \varphi$ is increasing.

Proof. (i) The continuity of φ follows the Maximum Theorem (see the Appendix). But, we give here an elementary proof.
Let (k_0^n) be a sequence of nonnegative numbers which converge to k_0. Let (k_1^n) be the associated sequence of maximizers, i.e., $\forall n, k_1^n = \varphi(k_0^n)$. Since $0 \leq k_1^n \leq f(k_0^n)$, the sequence (k_1^n) is in a bounded set. Let k_1 be a cluster point of this sequence. There exists a subsequence $(k_1^{n_k})$ which converges to k_1. Obviously, we have $0 \leq k_1 \leq f(k_0)$. Since $\forall k,\ V(k_0^{n_k}) = u(f(k_0^{n_k}) - k_1^{n_k}) + \beta V(k_1^{n_k})$, by taking the limits when k converges to infinity, one gets $V(k_0) = u(f(k_0) - k_1) + \beta V(k_1)$. This shows that $k_1 = \varphi(k_0)$. The cluster point is hence unique and the whole sequence (k_1^n) converges to k_1.
(ii) If $k_0 > 0$, then the optimal path $\mathbf{k}$ from k_0 satisfies $k_1 > 0$ and $f(k_0) - k_1 > 0$ (see Proposition 2.3.1). But $k_1 = \varphi(k_0)$. We have proved that the statement is true.
(iii) From Proposition 2.4.2, a feasible sequence $\mathbf{k}$ is optimal if, and only if,

$$\forall t, V(k_t) = u(f(k_t) - k_{t+1}) + \beta V(k_{t+1}).$$

This shows that $\forall t, k_{t+1} = \varphi(k_t)$, hence, $\forall t, k_t = \varphi^t(k_0)$.
(iv) Choose $k_0 < k_0'$. If $k_0 = 0$ then $\varphi(k_0) = 0$. Since $\varphi(k_0') > 0$, the statement is true in this case. So, assume $k_0 > 0$.
Assume the contrary, i.e., $\varphi(k_0) \geq \varphi(k_0')$. Write $k_1 = \varphi(k_0),\ k_2 = \varphi(k_1),\ k_1' = \varphi(k_0'), k_2' = \varphi(k_1')$. By the Euler equation we have:

$$u'(f(k_0) - k_1) = \beta u'(f(k_1) - k_2)f'(k_1),$$

$$u'(f(k_0') - k_1') = \beta u'(f(k_1') - k_2')f'(k_1').$$

From Proposition 2.4.1, $V'(k_1) = u'(f(k_1) - k_2)f'(k_1)$, and $V'(k_1') = u'(f(k_1') - k_2')f'(k_1')$. By the concavity of V, if $k_1 \geq k_1'$, then $V'(k_1) \leq V'(k_1')$, hence, $f(k_0) - k_1 \geq f(k_0') - k_1'$ since u' is decreasing. Since we assume $k_1 \geq k_1'$, this implies $k_0 \geq k_0'$, a contradiction with the choice of k_0 and k_0'. One concludes that $\varphi(k_0') > \varphi(k_0)$.
(v) We have $u'(f(k_0) - \varphi(k_0)) = \beta V'(\varphi(k_0))$. From statement (iv), if k_0 increases, then $\varphi(k_0)$ increases and hence, $V'(\varphi(k_0))$ decreases. That implies that $f(k_0) - \varphi(k_0)$ increases. □

2.4.4 Dynamic Properties of the Optimal Path

In this section, we show that the optimal path is monotonic and converges to an optimal steady state. A point k^s is an *optimal steady state* if the stationary sequence $(k^s, k^s, ..., k^s, ...)$ solves the problem

$$\max \sum_{t=0}^{+\infty} \beta^t u(f(k_t) - k_{t+1})$$

under the constraints:

$$\forall t \geq 0,\ 0 \leq k_{t+1} \leq f(k_t),$$

and $k_0 = k^s$.
Another definition is : k^s is an *optimal steady state* if $k^s = \varphi(k^s)$.

Proposition 2.4.4. *(i) The optimal capital stocks path* $\mathbf{k}$ *from* k_0 *is monotonic.*
(ii) Assume $k_0 > 0$*. Then:*
(a) If $f'(0) \leq 1/\beta$*, then the optimal path* $\mathbf{k}$ *converges to zero.*
(b) If $f'(0) > 1/\beta$*, then the optimal path* $\mathbf{k}$ *converges to an optimal steady state* k^s *which is strictly positive and satisfies* $f'(k^s) = 1/\beta$*.*

Proof. (i) If $k_0 = k_1$ then $k_0 = \varphi(k_0)$, i.e., k_0 is an optimal steady state.
If $k_0 < k_1$, then $k_1 < k_2$, since $k_1 = \varphi(k_0)$, $k_2 = \varphi(k_1)$ and φ is increasing. By induction, $k_t < k_{t+1}, \forall t$.
The same argument holds when $k_0 > k_1$.
(ii) Let $\mathbf{k}$ be the optimal path from $k_0 > 0$. We know that $\forall t, k_t \leq \max\{k_0, \bar{k}\}$ where $\bar{k}$ satisfies $f(\bar{k}) = \bar{k}$. Since $\mathbf{k}$ is monotonic, it must converge to some $k^s \geq 0$. Since φ is continuous and since $\forall t, k_{t+1} = \varphi(k_t)$, we have $k^s = \varphi(k^s)$, i.e. k^s is an optimal steady state. If $k^s > 0$, then $f(k^s) - k^s = f(k^s) - \varphi(k^s) > 0$ (see Proposition 2.4.3). Consider the Euler Equation:

$$\forall t, u'(f(k_t) - k_{t+1}) = \beta u'(f(k_{t+1}) - k_{t+2}) f'(k_{t+1}).$$

If $k^s > 0$, by taking the limits of $\mathbf{k}$ in the Euler equation, we obtain $f'(k^s) = 1/\beta$.

(a) If $f'(0) \leq 1/\beta$, such a $k^s > 0$ does not exist because f' is decreasing. In this case, one must conclude that $k^s = 0$, i.e., the sequence $\mathbf{k}$ converges to zero.
(b) Assume now that $f'(0) > 1/\beta$. We claim that the sequence $\mathbf{k}$ does not converge to zero. Indeed, if it converges to zero, then from the Euler equation, for t large enough, we will have $u'(f(k_t) - k_{t+1}) > u'(f(k_{t+1}) - k_{t+2})$, i.e., $f(k_{t+1}) - k_{t+2} > f(k_t) - k_{t+1}$. That means there exists some T such that

$$\forall t \geq T, \;\; f(k_t) - k_{t+1} > f(k_T) - k_{T+1} > 0.$$

But, if (k_t) converges to zero, the term $f(k_t) - k_{t+1}$ will converge also to zero, and we obtain a contradiction. Hence, the sequence $\mathbf{k}$ must converge to some $k^s > 0$ which satisfies $f'(k^s) = 1/\beta$. □

2.4.5 Mangasarian Lemma

We know that when $k_0 > 0$, the optimal path $\mathbf{k}$ satisfies the Euler equation. If we know k_0 and k_1, the Euler equation gives the value of $k_2, k_3, \ldots$. But, in our model, we just have k_0 as initial data. Therefore, there are many solutions to the Euler equation. We want to find among them the optimal path $\mathbf{k}$. In the following lemma, we give a second condition which ensures that a solution to the Euler equation is actually the optimal solution. The following Lemma has been proved by Weitzman [85].

Lemma 2.4.1. **[Mangasarian Lemma]** *In the Ramsey model, a feasible path* $\mathbf{k}$ *from* k_0 *is optimal if, and only if, it satisfies*
(i) **Euler Equation**

$$\forall t, u'(f(k_t) - k_{t+1}) = \beta u'(f(k_{t+1}) - k_{t+2}) f'(k_{t+1}) \qquad (\mathcal{E})$$

(ii) **Tranversality Condition**

$$\lim_{t \to +\infty} \beta^t u'(f(k_t) - k_{t+1}) k_{t+1} = 0. \qquad (\mathcal{T}')$$

Proof. (i) If $\mathbf{k}$ is optimal from k_0, then it satisfies Euler equation (Proposition 2.3.1) and $\forall t, k_t \leq \max\{k_0, \bar{k}\}$. We have by the concavity of V:

$$\beta^{t+1}V(k_{t+1}) - \beta^{t+1}V(0) \geq \beta^{t+1}V'(k_{t+1})k_{t+1}.$$

From the transversality condition $(\mathcal{T})$, we have

$$\lim_{t\to+\infty} \beta^{t+1}V(k_{t+1}) = 0.$$

Since $V'(k_{t+1}) > 0, \forall t$, we have therefore,

$$\lim_{t\to+\infty} \beta^{t+1}V'(k_{t+1})k_{t+1} = 0.$$

But $V'(k_{t+1}) = u'(f(k_{t+1}) - k_{t+2})f'(k_{t+1})$ (Proposition 2.4.1). From the Euler equation, we have $\beta u'(f(k_{t+1})-k_{t+2})f'(k_{t+1}) = u'(f(k_t)-k_{t+1})$. Therefore, $\lim_{t\to+\infty} \beta^t u'(f(k_t)-k_{t+1})k_{t+1} = 0$. We obtain the transversality condition for the optimal sequence $\mathbf{k}$.

(ii) Conversely, let $\mathbf{k} \in \Pi(k_0)$ satisfy the Euler equation $(\mathcal{E})$ and the Transversality condition $(\mathcal{T})$. Let $\mathbf{k}'$ be another feasible path from k_0. Define

$$\Delta_T = \sum_{t=0}^{T} \beta^t u(f(k_t) - k_{t+1}) - \sum_{t=0}^{T} \beta^t u(f(k'_t) - k'_{t+1})$$

$$= u(f(k_0)-k_1)-u(f(k_0)-k'_1)+\beta(u(f(k_1)-k_2)-u(f(k'_1)-k'_2))+\ldots$$
$$+\beta^T(u(f(k_T) - k_{T+1}) - u(f(k'_T) - k'_{T+1})).$$

Using the concavity of u and f yields:

$$\Delta_T \geq u'(f(k_0) - k_1)(k'_1 - k_1) + \beta u'(f(k_1) - k_2)f'(k_1)(k_1 - k'_1) + \ldots +$$
$$\beta^{T-1}u'(f(k_{T-1} - k_T)(k'_T - k_T) + \beta^T u'(f(k_T) - k_{T+1})f'(k_T)(k_T - k'_T)$$
$$+\beta^T u'(f(k_T) - k_{T+1})(k'_{T+1} - k_{T+1}).$$

Apply the Euler equalities. Then

$$\Delta_T \geq \beta^T u'(f(k_T)-k_{T+1})(k'_{T+1}-k_{T+1}) \geq -\beta^T u'(f(k_T)-k_{T+1})k_{T+1}.$$

Let $T \to +\infty$. Then, $\lim_{T\to+\infty} \Delta_T \geq 0$. That implies:

$$\sum_{t=0}^{+\infty} \beta^t u(f(k_t) - k_{t+1}) \geq \sum_{t=0}^{+\infty} \beta^t u(f(k'_t) - k'_{t+1}).$$

In other words, the sequence $\mathbf{k}$ is optimal. □

Remark 2.4.2. How may be interpreted the tranversality condition $(\mathcal{T}')$? In Chapter 6, we will show that with the paths of optimal consumptions and capital stocks can be associated a sequence of prices such that, together, they constitute a competitive equilibrium of an intertemporal economy. The sequence of equilibrium prices will be $\forall t, p_t = \beta^t u'(f(k_t) - k_{t+1})$. The transversality condition means that the value of the capital stock on hand at period $t+1$ (but produced at period t) converges to zero in the long term.

2.4.6 On the Continuity of the Value Function and of the Optimal Policy with respect to the Discount Factor and the initial Capital Stock

Let $0 < \beta < 1$ and k_0 be given. We will now write $V(\beta, k_0)$ instead of $V(k_0)$. Let $\{(\beta_n, k_0^n)\}$ be a sequence with $k_0^n \geq 0, \forall n$, converging to (β, k_0). We want to prove that $V(\beta_n, k_0^n)$ converges to $V(\beta, k_0)$. Let $A > \max\{k_0, \bar{k}\}$ where $\bar{k} = f(\bar{k})$. For n large enough, we have $0 \leq k_0^n < A$. Let $\mathbf{k}$ be the optimal path associated with (β, k_0) and $\mathbf{k^n}$ be the optimal path associated with (β_n, k_0^n).
We know that $\forall t, \forall n, 0 \leq k_t^n \leq A$. We can therefore assume that the sequence $\mathbf{k^n}$ converges to some $\mathbf{k'} \in \Pi(k_0)$, in the product topology, when $n \to +\infty$.
Let $T > 0$. We have

$$V(\beta_n, k_0^n) \leq \sum_{t=0}^{T} \beta_n^t u(f(k_t^n) - k_{t+1}^n) + \frac{\beta_n^T}{1-\beta_n} u(f(A)).$$

Hence

$$\limsup_n V(\beta_n, k_0^n) \leq \sum_{t=0}^{T} \beta^t u(f(k_t') - k_{t+1}') + \frac{\beta^T}{1-\beta} u(f(A)).$$

Let $T \to +\infty$. Then

$$\limsup_n V(\beta_n, k_0^n) \leq \sum_{t=0}^{+\infty} \beta^t u(f(k_t') - k_{t+1}') \leq V(\beta, k_0).$$

We will prove that $\liminf_n V(\beta_n, k_0^n) \geq V(\beta, k_0)$.

First, when $k_0 = 0$, then $k_t = 0, \forall t$ and $V(\beta_n, k_0^n) \geq 0 = V(\beta, 0)$. Thus

$$\liminf_n V(\beta_n, k_0^n) \geq V(\beta, k_0).$$

Now assume $k_0 > 0$. Because of the Inada condition on the utility function u we have $0 < k_1 < f(k_0)$. Therefore, for n large enough,

$$0 < k_1 < f(k_0^n),$$

and $(k_0^n, k_1, ..., k_t, ...) \in \Pi(k_0^n)$. Hence

$$V(\beta_n, k_0^n) \geq u(f(k_0^n) - k_1) + \sum_{t \geq 1} \beta_n^t u(f(k_t) - k_{t+1}).$$

That implies

$$\liminf_n V(\beta_n, k_0^n) \geq V(\beta, k_0).$$

We have proved that V is continuous with respect to (β, k_0).
Given β, denote, as for the Value function, by $\varphi(\beta, k_0)$ the value of the optimal policy at k_0, instead of $\varphi(k_0)$. Bellman equation gives:

$$V(\beta_n, k_0^n) = u(f(k_0^n) - k_1^n) + \beta_n V(\beta_n, k_1^n)$$

where $k_1^n = \varphi(\beta_n, k_0^n)$.
Let $n \to +\infty$. We have

$$V(\beta, k_0) = u(f(k_0) - k_1') + \beta V(\beta, k_1')$$

where k_1' is an accumulation point of $\{k_1^n\}$ (see the Appendix). Since u is strictly concave, $k_1' = \varphi(\beta, k_0)$ and $\varphi(\beta_n, k_0^n) \to \varphi(\beta, k_0)$.
We summarize these results in

Proposition 2.4.5. *The Value function and the optimal policy are continuous with respect to* (β, k_0).

3. An Aggregated Optimal Growth Model with a Convex-Concave Production Function

In Chapter 2, the concavity of the production function plays a crucial role.Two key results rely on this property. First, the optimal capital stocks converge monotonically towards an optimal steady state. Second, the Value function is a concave differentiable function of the initial capital stock.

Clark [23] initiated an analysis, completed later by Majumdar and Mitra [51], for a problem which was equivalent to an optimal growth model with a linear utility function and a convex-concave production function. What Clark speculated and Majumdar and Mitra proved is that for an interval of values of the discount factor, the economy exhibit a "trap". In other words, there exists a critical value of the initial stock below which the optimal path converges to zero.

Majumdar and Mitra [52] extended their work to a growth problem where the utility function is strictly concave and the production function has the same property as before. They demonstrated that, if the discount factor is high, then the optimal path converges to a steady state, while it converges to zero when the discount factor is low. However, these authors did not give definitive results when the discount factor is in the intermediate range.

Dechert and Nishimura [29] gave a complete characterization of the optimal growth path when the utility function is strictly concave and the production function is convex-concave. They used Bellman equation to prove the monotonicity of the optimal paths. In particular, they proved, for the case of intermediate discounting, that there exists a critical value of the initial capital stock below which any optimal path converges to the origin, and beyond which, all optimal paths converge to a steady state. They also proved that the Value function has left and right derivatives everywhere. Actually, they had stronger results. The Value function is differentiable almost everywhere in $\mathbb{R}_+$ and from almost everywhere in $\mathbb{R}_+$, there exists a unique optimal path. This chapter of our book is based on the paper by Dechert and Nishimura [29].

Let us mention that the model of Dechert and Nishimura has been adapted, in continuous time horizon, for a developing country which faces foreign debt, by Askenazy and Le Van [6], and in discrete time horizon setting for a developing country which faces debt, corruption and R&D, by Dimaria and Le Van [32].

A generalization of the model of Dechert and Nishimura [29] is given by Amir [4] who considers a multi-sector nonclassical optimal growth model. He uses lattice programming and Topkis' theorem to prove that the optimal correspondence is increasing and hence that the optimal paths are monotonic.

Concerning the differentiability of the Value function, Dechert and Nishimura [29]and Amir [4] exploit the fact that the optimal correspondence is increasing to prove that the left derivative of the Value function is less or equal to its right derivative. Askri and Le Van [7] obtain the same result in a multi-sector nonclassical optimal growth model using Clarke's [24] generalized gradients instead of lattice programming.

3.1 The Model

The model differs from the standard Ramsey Model by the assumptions on the production function. Formally, the model is stated as below:

$$\max \sum_{t=0}^{\infty} \beta^t u(c_t)$$

under the constraints:

$$\forall t, c_t + k_{t+1} \leq f(k_t),$$

$$\forall t, c_t \geq 0, k_t \geq 0,$$

and $k_0 \geq 0$ is given.

As in Chapter 2, c_t is the consumption at period t, while k_t is the capital stock on hand at date t.

3.1.1 The Assumptions

The following assumptions will be maintained throughout this chapter.

H0 $0 < \beta < 1$.
H1 The function $u : \mathbb{R}_+ \to \mathbb{R}_+$, is twice continuously differentiable and satisfies $u(0) = 0$. Moreover, its derivatives satisfy $u' > 0$ (strictly increasing) and $u'' < 0$ (strictly concave).
H2 Inada Condition : $u'(0) = +\infty$.
H3 The function $f : \mathbb{R}_+ \to \mathbb{R}_+$ is twice continuously differentiable and satisfies $f(0) = 0$. Its derivatives satisfy $f' > 0$ (strictly increasing).
H4 There is a point k_I such that $f''(k) < 0$ if $k > k_I$, and $f''(k) > 0$ if $k < k_I$.
H5 There exists a point $k_{max} > k_I$ such that $f(k_{max}) = k_{max}$ and $f(k) < k$ if $k > k_{max}$.
Assumption **H4** implies that f is strictly convex when $k < k_I$ and strictly concave when $k > k_I$. It says also that the function f' has a unique maximizer which is k_I.

3.1.2 Feasible Paths

Let us recall the following definitions which have been introduced in Chapter 2.

A sequence $(c_t, k_t)_{t=0,\dots,+\infty}$ is feasible from k_0 if it satisfies:

$$\forall t, c_t + k_{t+1} \leq f(k_t),$$

$$c_t \geq 0, \; k_t \geq 0,$$

and k_0 is given.

A sequence of consumptions $(c_t)_{t=0,\dots,+\infty}$ is feasible from k_0 if there exists a sequence of capital stocks $(k_t)_{t=0,\dots,+\infty}$ such that, together, they satisfy the previous relations.

Since the utility function is strictly increasing, at the optimum, the constraints will be binding. Thus, the initial optimal growth model is equivalent to the following problem:

$$\max \sum_{t=0}^{+\infty} \beta^t u(f(k_t) - k_{t+1})$$

under the constraints:

$$\forall t,\ 0 \leq k_{t+1} \leq f(k_t),$$

and $k_0 \geq 0$ is given.

A sequence of capital stocks $(k_t)_{t=0,\ldots,+\infty}$ is feasible from $k_0 \geq 0$ if we have

$$\forall t,\ 0 \leq k_{t+1} \leq f(k_t).$$

We will denote by $\mathbf{k}$ a feasible sequence of capital stocks from k_0 and by $\Pi(k_0)$ the sets of feasible capital $\mathbf{k}$.

3.2 Existence of Optimal Paths

To prove the existence of optimal capital and optimal consumption paths, we show that the initial optimization problem turns out to be a maximization problem of a continuous function on a compact set. We use, for this purpose, the product topology (for a definition, see the Appendix) defined on the space of infinite sequences of real numbers.

The main result of this section is:

Proposition 3.2.1. *There exists an optimal path* $\mathbf{k}$. *The associated optimal consumption path* $\mathbf{c}$ *is*

$$c_t = f(k_t) - k_{t+1}, \forall t.$$

Proof. Use the proof of Proposition 2.2.1 □

3.3 Properties of the Optimal Paths

In this section, we shall first prove that the solution $(\mathbf{k}, \mathbf{c})$ to the optimal growth model satisfies $\forall t, c_t > 0$, $k_t > 0$. This follows from Inada's condition on u. We then show that the optimal capital stocks satisfy the **Euler equation**.

Proposition 3.3.1. *(i) Every solution* $(\mathbf{k}, \mathbf{c})$ *to the optimal growth model satisfies* $\forall t, c_t > 0,\ k_t > 0,$ *if* $k_0 > 0$.
(ii) When $k_0 > 0$*, any solution satisfies the Euler equation:*

$$\forall t, u'(f(k_t) - k_{t+1}) = \beta u'(f(k_{t+1}) - k_{t+2}) f'(k_{t+1}) \qquad (\mathcal{E})$$

Proof. It is the same as in Proposition 2.3.1 □

Remark 3.3.1. Even though the utility function is strictly concave, we cannot claim that the solution is unique, as for the standard Ramsey Model, because the function f is not concave.

3.4 Value Function - Bellman Equation - Optimal Policy

As in the standard Ramsey Model, the Value function V is defined by:

$$\forall k_0 \geq 0, V(k_0) = \max \sum_{t=0}^{+\infty} \beta^t u(f(k_t) - k_{t+1})$$

under the constraints:

$$\forall t,\ 0 \leq k_{t+1} \leq f(k_t),$$

and $k_0 \geq 0$ is given.

3.4.1 Some Properties of the Value Function

We will show that the Value function is continuous and strictly increasing. The Value function V, in the standard Ramsey Model of Chapter 2, is concave and finite valued on $[0, +\infty[$. Since any concave function is continuous in the interior of its domain, we just had to prove that V is continuous at the origin. Here, since the production function is not concave, the Value function is also not concave. We will give below a direct proof.

Proposition 3.4.1. *The Value function satisfies:*
(i) $V(0) = 0; V(k_0) > 0$, *if* $k_0 > 0$. *If* $\mathbf{k}$ *is the optimal path, then* $V(k_0) = \sum_{t=0}^{+\infty} \beta^t u(f(k_t) - k_{t+1})$.
(ii) V *is continuous.*
(iii) V *is strictly increasing.*
(iv) $\forall \mathbf{k} \in \Pi(k_0)$, $\lim_{t\to+\infty} \beta^t V(k_t) = 0$.

Proof. (i) The proof of this statement is straightforward.
(ii) To prove the continuity of V: take a sequence $\{k_0^n\}$ converging to k_0. Let $\mathbf{k^n}, \mathbf{k}$ be respectively solutions associated with k_0^n and k_0. We know that if $\mathbf{k} \in \Pi(k_0)$, then $\forall t,\ 0 \le k_t \le \max\{k_0, k_{max}\}$. Hence, for any n large enough, say $n \ge N_1$, we will have $0 \le k_t^n \le \max\{2k_0, k_{max}\}, \forall t$. Let $b = \max\{2k_0, k_{max}\}$. For $n \ge N_1$,:

$$|\sum_{t=0}^{+\infty} \beta^t u(f(k_t^n) - k_{t+1}^n) - \sum_{t=0}^{+\infty} \beta^t u(f(k_t) - k_{t+1})|$$

$$\le \sum_{t=0}^{T} \beta^t |u(f(k_t^n) - k_{t+1}^n) - u(f(k_t) - k_{t+1})| + 2u(f(b))\beta^T/(1-\beta).$$

Let $\varepsilon > 0$ be given. Choose T such that $2u(f(b))\beta^T/(1-\beta) < \varepsilon/2$. Fix such a T. Then there exists $N_2 \ge N_1$ such that,

$$\forall n \ge N_2, \sum_{t=0}^{T} \beta^t |u(f(k_t^n) - k_{t+1}^n) - u(f(k_t) - k_{t+1})| < \varepsilon/2,$$

because, since u and f are continuous, $|u(f(k_t^n) - k_{t+1}^n) - u(f(k_t) - k_{t+1})|$ converges to zero for $t = 0, 1, 2, .., T$. We have proved the continuity of V.
(iii) See the proof of Proposition 2.4.1
(iv) We know that for any $\mathbf{k} \in \Pi(k_0)$, we have $\forall t, 0 \le k_t \le B = \max\{k_0, k_{max}\}$. Since V is increasing, we have $0 \le V(k_t) \le V(B), \forall t$. Hence, $\lim_{T\to+\infty} \beta^T V(k_T) = 0$. □

3.4.2 Bellman Equation

Proposition 3.4.2. *(i) The value function* V *satisfies the Bellman equation:*

$$\forall k_0 \geq 0, V(k_0) = \max\{u(f(k_0) - k) + \beta V(k) : 0 \leq k \leq f(k_0)\} \quad (\mathcal{B})$$

(ii) The Value function V is the unique solution to the Bellman equation, which is continuous and which satisfies the transversality condition:

$$\forall k_0 \geq 0,\ \forall \mathbf{k} \in \Pi(k_0),\ \lim_{T \to +\infty} \beta^T V(k_T) = 0 \quad (\mathcal{T})$$

(iii) A sequence $\mathbf{k} \in \Pi(k_0)$ is an optimal solution if, and only if, it satisfies:

$$\forall t,\ V(k_t) = u(f(k_t) - k_{t+1}) + \beta V(k_{t+1}).$$

Proof. The proof is the same as for Proposition 2.4.2 □

3.4.3 Optimal Correspondence

Let $k_0 \in \mathbb{R}_+$ and let $G(k_0) = \text{Argmax}\ \{u(f(k_0) - y) + \beta V(y) : y \in [0, f(k_0)]\}$. In Chapter 2, the functions u and f being strictly concave, the correspondence G is single-valued. Here, the function f is not concave. Hence, G may be multivalued. For this reason, we call it an optimal correspondence.

Proposition 3.4.3. *The correspondence G satisfies the following properties:*
(i) $G(0)=\{0\}$.
(ii) If $k_0 > 0$ and if $k_1 \in G(k_0)$, then $0 < k_1 < f(k_0)$.
(iii) Let (k_0^n) be a sequence of nonnegative numbers which converge to k_0, and let (k_1^n) be a sequence which satisfies $k_1^n \in G(k_0^n), \forall n$. Then there exists a converging subsequence $(k_1^{n_\nu})$ to $k_1 \in G(k_0)$. In particular, $G(k_0)$ is compact for every $k_0 \in \mathbb{R}_+$.
(iv) A sequence $\mathbf{k} \in \Pi(k_0)$ is optimal from k_0 if, and only if, it satisfies $\forall t \geq 1, k_t \in G(k_{t-1})$.
(v) The correspondence G is increasing, i.e., if $k_0 < k_0'$, $k_1 \in G(k_0)$, $k_1' \in G(k_0')$, then $k_1 < k_1'$.

Proof. (i) It is obvious that $G(0) = \{0\}$.
(ii) Let $k_1 \in G(k_0)$. If $(k_2, k_3, ..., k_t, ...)$ is an optimal path from k_1, then, obviously, $(k_1, k_2, ...k_t, ...)$ is an optimal path from k_0. From Proposition 3.3.1, we have $0 < k_1 < f(k_0)$.

(iii)Let (k_0^n) be a sequence of nonnegative numbers which converge to k_0. We have: $\forall n,\ 0 \le k_1^n \le f(k_0^n)$ and $k_1^n \in G(k_0^n)$. Since f is continuous and since $k_0^n \to k_0$, the sequence (k_1^n) is contained in a fixed interval $[0, \hat{A}]$. Consequently, there exists a subsequence $(k_1^{n_\nu})$ which converges to $k_1 \in [0, f(k_0)]$. We want to show that

$$u(f(k_0) - k_1) + \beta V(k_1) \ge u(f(k_0) - y) + \beta V(y), \forall y \in [0, f(k_0)].$$

Take $y \in [0, f(k_0)]$. If $y < f(k_0)$, then for ν large enough, we have $y \le f(k_0^{n_\nu})$. Hence,

$$u(f(k_0^{n_\nu}) - k_1^{n_\nu}) + \beta V(k_1^{n_\nu}) \ge u(f(k_0^{n_\nu}) - y) + \beta V(y),$$

and by taking the limits we obtain the desired inequality.
If $y = f(k_0)$, then $y = \lim_\nu y^{n_\nu}$ where $y^{n_\nu} = f(k_0^{n_\nu})$. Then

$$u(f(k_0^{n_\nu}) - k_1^{n_\nu}) + \beta V(k_1^{n_\nu}) \ge u(f(k_0^{n_\nu}) - y^{n_\nu}) + \beta V(y^{n_\nu}).$$

Again, take the limits to obtain the desired inequality.
Now, if $\{k_1^n\} \subset G(k_0)$ is a sequence which converges to some k_1, then, by applying the same argument as above, we obtain $k_1 \in G(k_0)$. That proves that $G(k_0)$ is closed. Since it bounded, it is therefore compact.
(iv) The argument is the same as for statement (iii) of Proposition 2.4.3
(v) Assume first that $k_1 > k_1'$. Then $f(k_0') > f(k_0) \ge k_1 > k_1'$. We have successively

$$V(k_0) = u(f(k_0) - k_1) + \beta V(k_1) \ge u(f(k_0) - k_1') + \beta V(k_1'),$$

and

$$V(k_0') = u(f(k_0') - k_1') + \beta V(k_1') \ge u(f(k_0') - k_1) + \beta V(k_1).$$

Hence, by adding the two inequalities, we get:

$$u(f(k_0) - k_1) + u(f(k_0') - k_1') \ge u(f(k_0) - k_1') + u(f(k_0') - k_1),$$

or equivalently,

$$0 \ge [u(f(k_0) - k_1') - u(f(k_0) - k_1)] + [u(f(k_0') - k_1) - u(f(k_0') - k_1')].$$

As

$$u(f(k_0) - k'_1) - u(f(k_0) - k_1) = \int_{k_1}^{k'_1} -u'(f(k_0) - x)dx,$$

$$u(f(k'_0) - k_1) - u(f(k'_0) - k'_1) = \int_{k'_1}^{k_1} -u'(f(k'_0) - x)dx,$$

we obtain

$$0 \geq \int_{k'_1}^{k_1} [u'(f(k_0) - x) - u'(f(k'_0) - x)]dx.$$

Since $f(k_0) - x < f(k'_0) - x$ for every $x \geq 0$, and u' is strictly decreasing, we have

$$\int_{k'_1}^{k_1} [u'(f(k_0) - x) - u'(f(k'_0) - x)]dx > 0$$

and hence, a contradiction. Thus, if $k_0 < k'_0$, then $k'_1 \geq k_1$. If $k'_1 = k_1$, take $k_2 \in G(k_1) = G(k'_1)$. We can write Euler equations:

$$u'(f(k_0) - k_1) = \beta u'(f(k_1) - k_2)f'(k_1),$$

$$u'(f(k'_0) - k_1) = \beta u'(f(k_1) - k_2)f'(k_1).$$

These relations imply that $k_0 = k'_0$ in contradiction with the initial assumption. □

Remark 3.4.1. The property (iii) of the correspondence G, in the previous proposition, is called upper- semi continuity. This result may be derived from the Maximum Theorem as one can see in Chapter 4 where we study a Multisector Optimal Growth Model.

3.4.4 On the Differentiability of the Value Function

In Chapter 2, the production function being concave, the Value function is differentiable at any $k_0 > 0$. When the production function is convex-concave, we will show that the Value function has left and right derivatives $V'_-(k_0), V'_+(k_0)$ at any $k_0 > 0$ fulfilling $V'_-(k_0) \leq V'_+(k_0)$. We will actually prove that the Value function is

differentiable almost everywhere. We will also show that the Value function is *differentiable* at a point $k_0 > 0$ iff there exists a *unique* optimal path from k_0. Hence, there exists a unique optimal path from almost everywhere.
The increasingness of the optimal correspondence G is crucial to prove that $V'_-(k_0) \leq V'_+(k_0)$. In multi-sector nonclassical optimal growth models, Amir [4] uses lattice programming and Topkis theorem to prove that the optimal policy is increasing. Again, this fact is exploited to prove that the left derivative of the Value function is less or equal to the right derivative of the Value function. Askri and Le Van [7] use Clarke [24] generalized gradients to obtain the same result. Their model is more general than Amir [4] model. It requires neither that the image by the technology correspondence of a sublattice is a sublattice, nor that the utility function is supermodular.

Proposition 3.4.4. *The Value function V has left and right derivatives at any $k_0 > 0$.*
More precisely,
(i) $V'_-(k_0) = u'(f(k_0) - k_1)f'(k_0)$ *where* $k_1 = \min\{x : x \in G(k_0)\}$ *and*
(ii) $V'_+(k_0) = u'(f(k_0) - k'_1)f'(k_0)$ *where* $k'_1 = \max\{x : x \in G(k_0)\}$.

Proof. Let $k_0 > 0$ and let $k_1 = \min\{x : x \in G(k_0)\}$. Let $\{k_0^n\}$ be a sequence of positive numbers converging to k_0 and satisfying $k_0^n < k_0$, $\forall n$. Let $k_n^1 \in G(k_0^n), \forall n$. We have :

$$V(k_0) = u(f(k_0) - k_1) + \beta V(k_1),$$

$$V(k_0^n) = u(f(k_0^n) - k_1^n) + \beta V(k_1^n),$$
$$0 < k_1 < f(k_0),$$

and by Proposition 3.4.3 (v)

$$\forall n, \ 0 < k_1^n < k_1 < f(k_0).$$

For n large enough, we have $0 < k_1 < f(k_0^n)$. Hence

$$V(k_0^n) \geq u(f(k_0^n) - k_1) + \beta V(k_1).$$

Therefore

$$V(k_0) - V(k_0^n) \leq u(f(k_0) - k_1) - u(f(k_0^n) - k_1)$$
$$\leq u'(f(k_0^n) - k_1)(f(k_0) - f(k_0^n)),$$

and

$$\limsup_{k_0^n \to k_0} \frac{V(k_0) - V(k_0^n)}{k_0 - k_0^n} \leq u'(f(k_0) - k_1)f'(k_0).$$

Similarly:

$$V(k_0) \geq u(f(k_0) - k_1^n) + \beta V(k_1^n),$$

and

$$V(k_0) - V(k_0^n) \geq u(f(k_0) - k_1^n) - u(f(k_0^n) - k_1^n)$$
$$\geq u'(f(k_0) - k_1^n)(f(k_0) - f(k_0^n)),$$

We can assume that $k_1^n \to k_1' \in G(k_0)$. Then

$$\liminf_{k_0^n \to k_0} \frac{V(k_0) - V(k_0^n)}{k_0 - k_0^n} \geq u'(f(k_0) - k_1')f'(k_0)$$

$$\geq u'(f(k_0) - k_1)f'(k_0), \text{ since } k_1' \geq k_1.$$

That means that the left derivative $V_-'(k_0)$ exists and is equal to $u'(f(k_0) - k_1)f'(k_0)$.

We now prove that $V_+'(k_0)$ exists and is equal to $u'(f(k_0) - k_1')f'(k_0)$ with $k_1' = \max\{x : x \in G(k_0)\}$.

Let $k_0'^n \to k_0$ and $k_0'^n > k_0, \forall n$. Let $k_1'^n \in G(k_0'^n), \forall n$. Since $0 < k_1' < f(k_0)$, then $0 < k_1' < f(k_0'^n)$ for every n large enough.

We claim that $k_1'^n < f(k_0)$ for any n large enough. Otherwise, there will be a subsequence $\{k_1'^\nu\}$ which satisfies $k_1'^\nu = f(k_0), \forall \nu$. We can assume that $k_1'^\nu$ converges to some $k_1'' \in G(k_0)$. But, from Proposition 3.4.3, we have $k_1'' < f(k_0)$ which contradicts the assumption.

An argument similar to the one used before for the left derivative leads to:

$$V_+'(k_0) = u'(f(k_0) - k_1')f'(k_0).$$

□

In the following corollary, we prove that the Value function is differentiable along an optimal path, and, that it is differentiable at an initial capital stock k_0 iff the optimal path from k_0 is unique. Moreover, the Value function is differentiable almost everywhere.

Corollary 3.4.1. *(i) Let* $\mathbf{k}$ *be an optimal path from* $k_0 > 0$. *The Value function* V *is differentiable at any* $k_t, t \geq 1$. *There exists a unique optimal path from any* $k_t, t \geq 1$.
(ii) The function V *is differentiable at* $k_0 > 0$ *iff there exists a unique optimal path from* k_0.
(iii) The Value function V *is differentiable almost everywhere. Equivalently, there exists a unique optimal path from almost any initial capital stock .*

Proof. (i) From the Euler equation, we have $u'(f(k_0) - k_1) = \beta u'(f(k_1) - k_2) f'(k_1)$. Hence k_2 is unique. By induction k_t is unique for $t \geq 2$. Since k_2 is unique, we have $V'_+(k_1) = V'_-(k_1) = V'(k_1)$, and for any $t \geq 2$, $V'_+(k_t) = V'_-(k_t) = V'(k_t)$.
(ii) If V is differentiable at k_0, we have $V'(k_0) = u'(f(k_0) - k_1) f'(k_0)$. Hence, k_1 is unique. From (i), $k_2, ..., k_t, ...$ are unique.
Conversely, if the optimal path is unique, then there exists a unique $k_1 \in G(k_0)$. From the previous proposition, $V'_+(k_0) = V'_-(k_0)$.
(iii) Let $\theta_1(k) = \max\{x : x \in G(k)\}$ and $\theta_2(k) = \min\{x : x \in G(k)\}$. The functions θ_1, θ_2 are increasing by Proposition 3.4.3 (v). Thus, they are continuous almost everywhere (see Appendix).
We will prove that the set of points $k > 0$ where θ_1 is continuous equals the set of points where θ_2 is continuous. Moreover, these functions coincide on this set.
First, let $k > 0$ be a point where θ_1 is continuous. Let $k' < k,\ k'' > k$ converge to k. We have:

$$\theta_1(\frac{k'^2}{k}) < \theta_2(k') < \theta_1(k') < \theta_1(k) < \theta_2(k'') < \theta_1(k'').$$

Hence:

$$\lim_{k' \to k} \theta_2(k') = \lim_{k'' \to k} \theta_2(k'') = \theta_1(k).$$

Since we have $\theta_2(k') < \theta_2(k) < \theta_2(k'')$, the function θ_2 is continuous at k and $\theta_2(k) = \theta_1(k)$.

Similarly, we prove that θ_1 is continuous where θ_2 is continuous and at this point $\theta_1 = \theta_2$.
We have proved that the correspondence G is single-valued almost everywhere, or equivalently, the Value function is differentiable almost everywhere. From statement (i), we conclude that there exists a unique optimal path from almost any initial capital stock. □

3.4.5 Dynamic Properties of the Optimal Paths

A point k^* is a *steady state* if either $k^* = 0$ or $f'(k^*) = 1/\beta$. In Chapter 2, we have shown that if $f'(0) \leq 1/\beta$, then the path of optimal capital stocks converges to zero, while if $f'(0) > 1/\beta$, then it converges to a steady state. Let us recall that the function f is defined by $f(k) = F(k) + (1-\delta)k$ where F is the production function and δ is the rate of depreciation of the capital. If we define the rate of interest r by $1/(1+r) = \beta$ (we discount the values of the utilities by the interest rate), then the condition $f'(0) \leq 1/\beta$ becomes $r + \delta \geq f'(0)$. Then when the cost of investment $(r+\delta)$ is high, the optimal capital stocks converge to zero. And when this cost is low, the optimal path converges to a steady state.

In presence of a convex-concave production function, there may be three steady states: the trivial one which is zero, and two other ones that we call low and high steady states. The stationary sequences, associated with these steady states, may or may not be optimal. There will be three regimes for the dynamics of the optimal path. When the cost of investment is low, the optimal path converges to the high steady state. There exists an intermediate range of values of the cost of investment for which the economy will face a "trap": there exists a critical value k_c such that, if $k_0 < k_c$, then the optimal path converges to zero, and if $k_0 > k_c$ then it converges to the high steady state. When the cost of investment is high, the dynamics will depend on the status of the high steady state. If it is not an optimal steady state, then the optimal path converges to zero, and if it is optimal, then, again, the economy will face a "trap".

Proposition 3.4.5. *Assume $k_0 > 0$. If $\mathbf{k}$ is an optimal path, then (i) it is monotonic and (ii) it converges to a steady state. Further-*

more, if the sequence **k** *is nondecreasing, then it converges to a strictly positive steady state.*

Proof. (i) the proof is the same as in Chapter 2.
(ii) (a) Assume first that the optimal path **k** is decreasing. Let k^* denote its limit. Since $k_{t+1} \in G(k_t)$ for every t, from Proposition 3.4.3 (iii), we have $k^* \in G(k^*)$. If $k^* > 0$, from Proposition 3.4.3 (ii), we have $k^* < f(k^*)$. Now, consider the Euler equation:

$$\forall t, u'(f(k_t) - k_{t+1}) = \beta u'(f(k_{t+1}) - k_{t+2}) f'(k_{t+1}).$$

By taking the limits, we obtain $f'(k^*) = 1/\beta$. Hence, either **k** converges to zero or to a point k^* with $f'(k^*) = 1/\beta$.
(b) Assume **k** non decreasing. Since $\forall t, 0 \leq k_t \leq \max\{k_0, k_{max}\}$, the sequence **k** must converge to some $k^* \geq k_0 > 0$. We must have $f'(k^*) = 1/\beta$. □

Theorem 3.4.1 (Mild Discounting).
Assume $f'(0) > 1/\beta$. *Then*
(i) There exists a unique $k^s \in]k_I, +\infty[$ *which satisfies* $f'(k^s) = 1/\beta$.
(ii) Any optimal path from $k_0 > 0$ *converges to* k^s.

Proof. (i) The function f' increases on $[0, k_I]$ and decreases on $[k_I, +\infty[$. As $f'(0) > 1/\beta$ and $f'(+\infty) \leq 1$, there exists a unique $k^s \in]k_I, +\infty[$ such that $f'(k^s) = 1/\beta$.
(ii) Let **k** be an optimal path. Consider the associated Euler equations:

$$\forall t, u'(f(k_t) - k_{t+1}) = \beta u'(f(k_{t+1}) - k_{t+2}) f'(k_{t+1}).$$

If $k_0 < k^s$ then **k** must be increasing. If not, it will be decreasing (it can not be stationary, since there exists a unique positive steady state) and converges to 0. In this case, $(f(k_t) - k_{t+1})$ converges also to zero, and $f'(k_t) \geq 1/\beta, \forall t$. From the Euler equations, one gets:

$$f(k_{t+1}) - k_{t+2} > f(k_t) - k_{t+1} > ... > f(k_0) - k_1 > 0,$$

yielding a contradiction.
Being increasing, the sequence **k** converges to k^s by Proposition 3.4.5.

Assume now $k_0 \geq k^s$. The sequence $\mathbf{k}$ is not increasing. If it were, it would, by Proposition 3.4.5, converge to a strictly positive steady state strictly higher than k^s, which is impossible. Hence, it is non increasing.
There is no t with $k_t < k^s$. If not, the sequence $(k_t, k_{t+1}, ...)$, which is optimal from k_t would be increasing, as we have shown before. And we obtain a contradiction with Proposition 3.4.5(i). Hence, $k_t \geq k^s, \forall t$, and converges to k^s. □

We will now examine the case of **Intermediate Discounting** $f'(0) < 1/\beta \leq \max_{k>0}\{f(k)/k\}$. The proofs of the results need many intermediate steps.

Lemma 3.4.1. *Let* $\mathbf{k}$ *be an optimal path from* k_0 *and let* $\mathbf{c}$ *be the associated optimal consumptions path, i.e.,* $\forall t, c_t = f(k_t) - k_{t+1}$. *Then*

$$f(k_0) - k_0 \leq \sum_{t=0}^{+\infty}(1-\beta)\beta^t c_t,$$

with strict inequality if $k_t \neq k_0$ *for some* t.

Proof. If $f(k_0) \leq k_0$, then the inequality obviously holds. Thus, assume $f(k_0) - k_0 > 0$. Then the sequence $\mathbf{k}'$ with $k'_t = k_0, \forall t$ is feasible. The associated feasible consumptions are $c'_t = f(k_0) - k_0, \forall t$.
Assume $f(k_0) - k_0 > \sum_{t=0}^{+\infty}(1-\beta)\beta^t c_t$. From Jensen's inequality; we get:

$$u(f(k_0) - k_0) > u(\sum_{t=0}^{+\infty}(1-\beta)\beta^t c_t) \geq (1-\beta)\sum_{t=0}^{+\infty}\beta^t u(c_t).$$

Divide the previous inequalities by $(1-\beta)$. We obtain

$$\sum_{t=0}^{+\infty}\beta^t u(f(k_0) - k_0) = \sum_{t=0}^{+\infty}\beta^t u(c'_t) > \sum_{t=0}^{+\infty}\beta^t u(c_t),$$

Contradicting the optimality of $\mathbf{c}$.
Hence $f(k_0) - k_0 \leq \sum_{t=0}^{+\infty}(1-\beta)\beta^t c_t$.

Assume now that $f(k_0) - k_0 = \sum_{t=0}^{+\infty}(1-\beta)\beta^t c_t$. We claim that $k_t = k_0, \forall t$. If not, let T be the first date such that $k_T \neq k_0$. We have $c_{T-1} = f(k_0) - k_T \neq f(k_0) - k_0$. There must be some t with $c_t \neq c_0$. If not, we would have

$$u(f(k_0) - k_0) = u(\sum_{t=0}^{+\infty}(1-\beta)\beta^t c_t) = u(c_0),$$

which implies $\forall t,\ c_t = c_0 = f(k_0) - k_0$ contradicting $c_{T-1} \neq f(k_0) - k_0$.
Hence, from Jensen's inequality, we get

$$u(f(k_0) - k_0) = u(\sum_{t=0}^{+\infty}(1-\beta)\beta^t c_t) > (1-\beta)\sum_{t=0}^{+\infty}\beta^t u(c_t),$$

which implies

$$\sum_{t=0}^{+\infty}\beta^t u(c'_t) > \sum_{t=0}^{+\infty}\beta^t u(c_t).$$

This contradicts the optimality of $\mathbf{c}$.
We have proved that, if $k_t \neq k_0$ for some t, then $f(k_0) - k_0 < \sum_{t=0}^{+\infty}(1-\beta)\beta^t c_t$. □

Lemma 3.4.2. *We have*
(i) $f'(k_I) > 1$.
(ii) The function $\frac{f(k)}{k}$ *has a unique maximizer* a *which belongs to* $]k_I, k_{max}]$ *and satisfies*

$$f'(0) < \frac{f(a)}{a} = f'(a) < f'(k_I).$$

Proof. (i) Since the function f' is maximal at k_I, if $f'(k_I) \leq 1$, then $f'(k) < 1$ for any $k \neq k_I$. In that case, $\forall k > 0, f(k) = \int_0^k f'(x)dx < k$ and Assumption **H5** does not hold. Thus, $f'(k_I) > 1$.
(ii) Recall that, from **H5**, on $[k_{max}, +\infty[$, we have $\frac{f(x)}{x} < 1 = \frac{f(k_{max})}{k_{max}}$. On $[0, k_I]$, as f is convex, the function $\frac{f(x)}{x}$ is strictly increasing. Hence, any maximizer a of $\frac{f(x)}{x}$ belongs to $[k_I, k_{max}]$. From

the first order condition, it satisfies $\frac{f(a)}{a} = f'(a)$. Therefore, it differs from k_I, since $\frac{f(k_I)}{k_I} < f'(k_I)$ as f is strictly convex on $[0, k_I]$ (see Appendix). Moreover, the maximizer a is unique, for, if there were two maximizers $a_1 < a_2$, we would have $f'(a_1) = f'(a_2)$, contradicting the strict concavity of f on $[k_I, +\infty[$. Since f' is strictly decreasing on $[k_I, +\infty[$, we have $f'(a) < f'(k_I)$ or $\frac{f(a)}{a} < f'(k_I)$. □

Lemma 3.4.3. *Assume $f'(0) < 1/\beta < f'(k_I)$. Then there exist k_s and k^s which satisfy:*

$$0 < k_s < k_I < k^s, \quad f'(k_s) = f'(k^s) = 1/\beta.$$

Proof. Consider the function $\psi(k) = f(k) - \frac{1}{\beta}k$. Then ψ' increases for $k \in [0, k_I]$, and decreases for $k \in [k_I, +\infty[$. Since $f'(k_I) > 1/\beta, f'(0) < 1/\beta$, there exists a unique $k_s \in]0, k_I[$ such that $\psi'(k_s) = 0$, or equivalently, $f'(k_s) = 1/\beta$. Since ψ' is decreasing in $[k_I, +\infty[$, and since $\psi'(\infty) < 1 - 1/\beta < 0$, there exists a unique $k^s \in]k_I, +\infty[$ with $\psi'(k^s) = 0$, i.e., $f'(k^s) = 1/\beta$. □

From now on, we will call k^s, the *high steady state* and k_s, the *low steady state.*

From Lemma 3.4.1 and Lemma 3.4.3 we obtain the following corollaries.

Corollary 3.4.2. *Assume $f'(0) < 1/\beta \leq \max\{f(k)/k : k > 0\}$. Then k^s is an optimal steady state, i.e. the sequence $\mathbf{k}$ with $k_t = k^s, \forall t \geq 0$, is optimal from k^s. Moreover, it is the unique optimal path from k^s.*

Proof. Consider the function ψ in the proof of Lemma 3.4.3. Observe that k^s maximizes ψ over $\mathbb{R}_+$. Hence, if $\mathbf{k}$ is a non stationary feasible sequence from k^s, then $f(k_t) - \frac{1}{\beta}k_t \leq f(k^s) - \frac{1}{\beta}k^s, \forall t$, with a strict inequality if $k_t \neq k^s$. Thus,

$$\sum_{t=0}^{+\infty} \beta^t [f(k_t) - \frac{1}{\beta}k_t] < \frac{1}{1-\beta}[f(k^s) - \frac{1}{\beta}k^s].$$

But

$$\sum_{t=0}^{+\infty} \beta^t [f(k_t) - \frac{1}{\beta} k_t] = \sum_{t=0}^{+\infty} \beta^t [f(k_t) - k_{t+1}] - \frac{1}{\beta} k^s,$$

thus,

$$\sum_{t=0}^{+\infty} \beta^t [f(k_t) - k_{t+1}] < \frac{1}{1-\beta} [f(k^s) - k^s],$$

in contradiction with Lemma 3.4.1. This shows that k^s is an optimal steady state and that the sequence $\mathbf{k}$ with $k_t = k^s, \forall t$, is the unique optimal sequence from k^s. □

We will now exhibit a very interesting result: No optimal path converges to the low steady state k_s.

Corollary 3.4.3. *Assume $f'(0) < 1/\beta < f'(k_I)$. If $k_0 \neq k_s$, then no optimal path from k_0 converges to k_s.*

Proof. Consider the function ψ defined by $\psi(k) = f(k) - \frac{1}{\beta} k$. It is decreasing on $[0, k_s]$ and is increasing on $[k_s, k^s]$.
First, assume $k_0 < k_s$. If an optimal path $\mathbf{k}$ converges to k_s, then it is increasing and we have $\psi(k_t) < \psi(k_0), \forall t$, i.e.,

$$f(k_t) - \frac{1}{\beta} k_t < f(k_0) - \frac{1}{\beta} k_0, \forall t.$$

Summing over t as in the proof of Corollary 3.4.2, we obtain

$$\sum_{t=0}^{+\infty} \beta^t (f(k_t) - k_{t+1}) < \frac{f(k_0) - k_0}{1-\beta},$$

in contradiction with Lemma 3.4.1.
Now, assume $k_0 > k_s$. Let $\mathbf{k}$ be an optimal path from k_0 which converges to k_s. There exists some T such that for any $t \geq T$, we have $k_s \leq k_t < k^s$. Recall that ψ is increasing on $[k_s, k^s]$. The sequence $(k_T, k_{T+1}, ..., k_{T+t} ...)$ is optimal from k_T. Since it is decreasing, we have $\psi(k_t) < \psi(k_T), \forall t > T$, which leads to

$$\sum_{t=0}^{+\infty} \beta^t (f(k_{T+t}) - k_{T+t+1}) < \frac{f(k_T) - k_T}{1-\beta},$$

in contradiction with Lemma 3.4.1. □

In the following Proposition, we show that, if k_s is an optimal steady state, then it is a critical point. In other words, for any initial capital stock lower than k_s, all the optimal paths converge to zero, and for any initial capital stock larger than k_s, the optimal paths converge to the high steady state k^s.

Proposition 3.4.6. *Assume $f'(0) < 1/\beta < f'(k_I)$ and that k_s is an optimal steady state. Then*
(i) The sequence $\mathbf{k}$ *defined by $k_t = k_s, \forall t$, is the unique optimal sequence from k_s.*
(ii) If $k_0 > k_s$ and if $\mathbf{k}$ *is an optimal sequence from k_0, then it converges to k^s. If $k_0 < k_s$ then* $\mathbf{k}$ *converges to* 0.

Proof. (i) Let $\mathbf{k}$ be another optimal sequence from k_s and let $t \geq 1$ be the first date such that $k_t \neq k_s$. Write the Euler equation:

$$u'(f(k_s) - k_s) = \beta u'(f(k_s) - k_t)f'(k_s).$$

Since $f'(k_s) = 1/\beta$, we have a contradiction with $k_t \neq k_s$.

(ii) If $k_0 > k_s$, then $\mathbf{k}$ is increasing and converges to k^s (Proposition 3.4.5 and Corollary 3.4.3).
If $k_0 < k_s$, then the optimal consequence converges to zero (by Proposition 3.4.5 and Corollary 3.4.3). □

But k_s may *not* be an optimal steady state as it is proved in the following Proposition. Observe that if $f(k_s) \leq k_s$, then it is obvious that k_s is *not* an optimal steady state, since the optimum steady state consumption $c_s = f(k_s) - k_s$ could not be negative or equal to zero.

Proposition 3.4.7. *Assume $f'(0) < 1/\beta < \max\{f(k)/k : k > 0\}$. Assume $f(k_s) > k_s$. Let $c_s = f(k_s) - k_s$. If $(\frac{1}{\beta} - 1)u''(c_s) + \beta u'(c_s)f''(k_s) > 0$, then k_s is not an optimal steady state.*

Proof. If k_s is optimal, then we have $0 < k_s < f(k_s)$ by Proposition 3.4.3 (ii). Let $\varepsilon > 0$ be so small that $0 < k_s + \varepsilon < f(k_s)$.
Define $\varphi(\varepsilon) = u(c_s - \varepsilon) + \frac{\beta}{1-\beta}u(f(k_s + \varepsilon) - k_s - \varepsilon)$. We have $\varphi(0) = \frac{u(f(k_s)-k_s)}{1-\beta}$, $\varphi'(0) = 0$, and

$$\varphi''(0) = \frac{1}{1-\beta}[(\frac{1}{\beta} - 1)u''(c_s) + \beta u'(c_s)f''(k_s)].$$

Hence $\varphi''(0) > 0$. Therefore, $\varphi(\varepsilon) > \varphi(0)$ for ε small enough. But

$$V(k_s) = \varphi(0) \geq u(f(k_s) - k_s - \varepsilon) + \beta V(k_s + \varepsilon)$$

$$\geq u(f(k_s) - k_s - \varepsilon) + \frac{\beta}{1-\beta} u(f(k_s + \varepsilon) - k_s - \varepsilon) = \varphi(\varepsilon),$$

for every $\varepsilon > 0$ small enough: a contradiction. □

We will now prove that when the initial capital stock is low, the optimal paths converge to zero. Observe that this Lemma requires $f'(0) < 1/\beta$, a weaker condition than the one which characterizes the Intermediate Discounting.

Lemma 3.4.4. *Assume $f'(0) < 1/\beta$. Then there exists $0 < \hat{k}_0 < k_s$ such that for any $k_0 < \hat{k}_0$, all optimal paths from k_0 converge to zero.*

Proof. Assume the contrary. There exists a positive sequence $\{k_0^n\}$ converging to zero such that, for any n, there exists an optimal sequence $\mathbf{k^n}$ from k_0^n which is increasing and converges to k^s. Let T_n denote the first date such that:

$$t \leq T_n + 1 \Rightarrow k_t^n \leq k_s,$$

and

$$t > T_n + 1 \Rightarrow k_t^n > k_s.$$

Write the Euler equations up to T_n:

$$\forall t \leq T_n, u'(f(k_t^n) - k_{t+1}^n) = \beta u'(f(k_{t+1}^n - k_{t+2}^n) f'(k_{t+1}^n).$$

Since $f'(k) \leq 1/\beta$, if $k \leq k_s$, we have:

$$0 < f(k_{t+1}^n) - k_{t+2}^n \leq f(k_t^n) - k_{t+1}^n \leq \ldots \leq f(k_0^n) - k_1^n \leq f(k_0^n).$$

From these inequalities, we obtain that the sequence $\{f(k_{T_n+1}^n) - k_{T_n+2}^n\}$ converges to zero when n converges to infinity. Now, since the set $\{k_{T_n+1}^n\}_{n=1,..,+\infty}$ is bounded, there exists a subsequence $\{k_{T_{n^k}+1}^{n^k}\}$ which converges to some $\bar{k} \leq k_s$, and $\{k_{T_{n^k}+2}^{n^k}\}$ converges to some $\bar{k}'$. From Proposition 3.4.3 (iii), $\bar{k}' \in G(\bar{k})$. Since $\bar{k}' \geq k_s > 0$, we must have $\bar{k} > 0$, by the statement (i) of the same proposition. The statement (ii) of the same proposition implies that $f(\bar{k}) - \bar{k}' > 0$. But we also have:

$$0 = \lim_{n^k \to +\infty} [f(k^{n^k}_{T_{n^k+1}}) - k^{n^k}_{T_{n^k+2}})] = f(\bar{k}) - \bar{k}',$$

which is a contradiction.
Therefore, there must exists $0 < k_0 < k_s$ such that any optimal path from k_0 converges to 0. □

We now prove that, when the initial capital stock is sufficiently high, the optimal path will converge to the high steady state.

Lemma 3.4.5. *Assume $f'(0) < 1/\beta \leq \max\{f(k)/k : k > 0\}$. Let $\tilde{k}$ denote the unique point in $]k_s, k^s[$ which satisfies $f(\tilde{k}) = \frac{1}{\beta}\tilde{k}$.*
We have: If $k_0 \geq \tilde{k}$, then any optimal sequence from k_0 converges to k^s.

Proof. If $k_0 = k^s$, from corollary 3.4.2, there exists a unique optimal sequence from k_0 which is $k_t = k^s, \forall t$. So, first, assume $k_0 > k^s$. From Proposition 3.4.5 and corollary 3.4.2, if $\mathbf{k}$ is an optimal path from k_0, we have $k_t > k^s, \forall t$. If $\mathbf{k}$ is increasing, it will converge to a point k^* with $f'(k^*) = \frac{1}{\beta}$ which is impossible. Hence $\mathbf{k}$ is decreasing and converges to k^s.
Now consider the case where $k_0 \in [\tilde{k}, k^s[$. Let ψ be the function defined by $\psi(k) = f(k) - \frac{1}{\beta}k$. It is increasing on $]k_s, k^s[$. Thus, if the optimal sequence $\mathbf{k}$ from k_0 is decreasing then, we have

$$f(k_t) - \frac{1}{\beta}k_t < f(k_0) - \frac{1}{\beta}k_0, \forall t,$$

which implies

$$\sum_{t=0}^{+\infty} \beta^t (f(k_t) - k_{t+1}) < \frac{f(k_0) - k_0}{1 - \beta},$$

in contradiction with Lemma 3.4.1. Hence, $\mathbf{k}$ is increasing. Proposition 3.4.5 implies that $k_t < k^s, \forall t$. Consequently, $\mathbf{k}$ converges to k^s. □

We have just shown that, when the initial capital stock is larger to some value, then the optimal capital stocks converge to the high steady state k^s. We have also shown in Lemma 3.4.4 that, when the initial capital stock is small, the optimal paths converge to zero. The following theorem states that, for the case of intermediate discounting, the economy exhibits a "poverty trap", i.e. a critical value for the initial capital stock.

Theorem 3.4.2 (Intermediate Discounting. Existence of a critical value). *Assume* $f'(0) < 1/\beta < \max\{f(k)/k : k > 0\}$. *Consider the point* $\tilde{k}$ *defined in Corollary 3.4.5. There exists* $k_c \in]0, \tilde{k}]$ *such that every optimal capital path starting from* $k_0 > k_c$ *converges to the high steady state* k^s*, and every optimal capital path starting from* $k_0 < k_c$ *converges to zero.*

Proof. We have proved that the set of points $k_0 > 0$ such that every optimal path from k_0 converges to zero, and the set of $k_0 > 0$ such that every optimal path from k_0 converges to k^s are *non empty* (Lemma 3.4.4, Lemma 3.4.5).
Let $\underline{k}_c, \bar{k}_c$ denote respectively the supremum and the infimum of the sets. Obviously $0 < \underline{k}_c \leq \bar{k}_c$. Now suppose $\underline{k}_c < \bar{k}_c$. Choose k_0, k_0' which satisfy $\underline{k}_c < k_0 < k_0' < \bar{k}_c$. There exist an optimal sequence from k_0', say $\mathbf{k}'$, which converges to zero, and another optimal sequence from k_0, say $\mathbf{k}$, which converges to k^s. In this case, for some t, one has $k_t' < k_t$ and that contradicts Proposition 3.4.5. Hence $\underline{k}_c = \bar{k}_c$. Let $k_c = \underline{k}_c = \bar{k}_c$. By corollary 3.4.5, we have $k_c \leq \tilde{k}$. Since $k_c = \underline{k}_c$, we have k_c¿0. □

We have mentioned that with a convex-concave production function, the optimal paths may not be unique. We will show below that it is the case at the critical value k_c.

Proposition 3.4.8. *Assume* $f'(0) < 1/\beta < \max\{f(k)/k : k > 0\}$ *and that the critical value* k_c *differs from the low steady state* k_s*. Then there exists an optimal path from* k_c *which converges to* k^s*, and another one which converges to zero.*

Proof. Let $\mathbf{k_0^n}$ be a sequence of positive numbers which satisfy $\forall n, k_0^n > k_c$ and $\mathbf{k_0^n}$ converges to k_c when n converges to infinity.

Let $\mathbf{k^n}$ denote the optimal path from k_0^n. This optimal path must converge to k^s by the previous Theorem, and hence, is increasing. Let $\mathbf{x} \in \Pi(k_c)$. Since we have $x_1 \leq f(k_c) \leq f(k_0^n)$, the sequence $\mathbf{x}$ is a also feasible from k_0^n. Therefore,

$$\sum_{t=0}^{+\infty} \beta^t u(f(k_t^n)) - k_{t+1}^n) \geq u(f(k_0^n) - x_1) + \sum_{t=1}^{+\infty} \beta^t u(f(x_t) - x_{t+1}).$$

We know that $\forall t, \forall n, k_t^n \leq \max\{k_0^n, k_{max}\}$. Hence, the sequences $\mathbf{k^n}$ are in a fixed compact set for the product topology. One can assume that $\mathbf{k^n}$ converges to some $\mathbf{k} \in \Pi(k_c)$. By taking the limits in the previous inequality, we obtain:

$$\sum_{t=0}^{+\infty} \beta^t u(f(k_t) - k_{t+1}) \geq u(f(k_c) - x_1) + \sum_{t=1}^{+\infty} \beta^t u(f(x_t) - x_{t+1}).$$

with $k_0 = k_c$. Since $\mathbf{x}$ is arbitrary in $\Pi(k_c)$, the previous inequality implies that $\mathbf{k}$ is optimal from k_c. Since $\forall t, \forall n, k_{t+1}^n \geq k_t^n$, we have $\forall t, k_{t+1} \geq k_t$. If $\forall t, k_{t+1} = k_t = k_c$, we must have either $k_c = k^s$, or $k_c = k_s$. The first equality is impossible since $k_c \leq \tilde{k} < k^s$ (Theorem 3.4.2). The second inequality is impossible by assumption. Hence, the sequence $\mathbf{k}$ must be increasing and converges to k^s.

The same argument applies for a sequence $\mathbf{k_0^n}$ which converges to k_c and which satisfies $k_0^n < k_c, \forall n$. The limit of $\mathbf{k_0^n}$ will give an optimal path from k_c which converges to zero. □

Finally, we consider the case of **Strong Discounting**, i.e., $\frac{1}{\beta} > \max\{f(k)/k : k > 0\}$. In this case, if $f'(k_I) < \frac{1}{\beta}$, then k^s, k_s do not exist. The conclusion is immediate: any optimal path converges to zero. So, we will assume that $f'(k_I) \geq \frac{1}{\beta} > \max\{f(k)/k : k > 0\}$.

Theorem 3.4.3 (Strong Discounting). *Assume $f'(k_I) \geq \frac{1}{\beta} > \max\{f(k)/k : k > 0\}$.*
(i) If k^s is not an optimal path, then all optimal paths from k_0 converge to zero.
(ii) If k^s is an optimal steady state, then there exists a unique $k_c \in]0, k^s]$ such that all optimal paths from $k_0 < k_c$ will converge to zero, and all optimal paths from $k_0 > k_c$ will converge to k^s.

When $f'(k_I) = \frac{1}{\beta}$ *and* k^s *is an optimal steady state, we have* $k_c = k^s = k_s$.

Proof. Observe that $f'(0) < 1/\beta$ since $f'(0) < \max\{f(k)/k : k > 0\}$. From Lemma 3.4.3, there are two steady states k_s, k^s (which could be equal).
(i) Assume the contrary. In this case, there exists an optimal path $\mathbf{k}$ which does not converge to zero. But from Corollary 3.4.3, it cannot converge to k_s. Hence, it converges to k^s. Since $\forall t, k_{t+1} \in G(k_t)$, from Proposition 3.4.3, we will have $k^s \in G(k^s)$, i.e., k^s is an optimal steady state: a contradiction with the assumption.
(ii) If $k_0 > k^s$ then, for any optimal path $\mathbf{k}$, we will have $k_t > k^s, \forall t$, and $\mathbf{k}$ converges to k^s (if not it will converge to another steady state larger than k^s, which is impossible). The set of k_0 such that all the optimal paths from k_0 converge to k^s is therefore non empty. We can again use Lemma 3.4.4 to claim that the set of k_0 such that all optimal paths starting from k_0 converge to zero is non empty. Use now the argument of Theorem 3.4.2 to conclude.
When $f'(k_I) = \frac{1}{\beta}$, we have $k_s = k^s = k_I$ and the function ψ, defined by $\psi(k) = f(k) - \frac{1}{\beta}k$, is decreasing on $[0, +\infty[$.
Let $k_0 \in]0, k^s[$. If an optimal path $\mathbf{k}$ from k_0 converges to k^s, we will have $\psi(k_t) < \psi(k_0), \forall t$. That will lead to

$$\sum_{t=0}^{+\infty} \beta^t (f(k_t) - k_{t+1}) < \frac{f(k_0) - k_0}{1 - \beta},$$

in contradiction with Lemma 3.4.1. Hence $\mathbf{k}$ converges to zero.
Let $k_0 > k^s$. Since we assume that k^s is an optimal steady state, we will have $k_t > k^s, \forall t$. The optimal path must converge to k^s. We have shown that k^s is the critical value.

□

3.5 On the Mangasarian Lemma

We have seen in Chapter 2, that a sequence $\mathbf{k}$ feasible from $k_0 > 0$ is optimal if, and only if, it satisfies the conditions of the Mangasarian Lemma: (i) Euler equation and (ii) Transversality Condition

$$\lim_{t \to +\infty} \beta^t u'(f(k_t) - k_{t+1})k_{t+1} = 0.$$

These conditions are not sufficient when the production is not concave.
Indeed, consider the low steady state k_s. The stationary sequence $(k_s, k_s, ...)$ fulfills the conditions of the Mangasarian Lemma, but it may not be optimal from k_s (see Proposition 3.4.7).
Are these conditions necessary when a path is optimal? We have proved that any optimal path from $k_0 > 0$ satisfies the Euler equations. The following proposition gives a condition implying the necessity of the transversality condition for any optimal path from $k_0 > 0$.

Proposition 3.5.1. *Assume* $f'(0) \geq 1$. *Let* $\mathbf{k}$ *be an optimal path from* $k_0 > 0$. *Then* $\lim_{t \to +\infty} \beta^t u'(f(k_t) - k_{t+1})k_{t+1} = 0$.

Proof. If $\mathbf{k}$ converges to k^s, then the statement is true since $\beta < 1$ and $f(k^s) - k^s > 0$.
Consider the case where $\mathbf{k}$ converges to zero. Since the derivative f' of f is strictly increasing in a neighborhood of the origin, there exists T such that for any $t \geq T$, we have $f'(k_{t+1}) \geq 1$. From Euler equations, for any $t \geq T$, we have $\beta^T u'(f(k_T) - k_{T+1}) \geq \beta^{t+1} u'(f(k_{t+1}) - k_{t+2}) \geq 0$. Hence, for every $t \geq T$:

$$\beta^{t+1} u'(f(k_{t+1}) - k_{t+2})k_{t+1} \leq \beta^T u'(f(k_T) - k_{T+1})k_{t+1}.$$

The conclusion is immediate since $\{k_t\}$ converges to zero. □

Remark 3.5.1. We let the reader check that, as in Chapter 2, the Value function is continuous with respect to (β, k_0).
Write $V(\beta, k_0)$ for the Value function and $G(\beta, k_0)$ for the optimal correspondence. Then:

$$G(\beta, k_0) = Argmax\{u(f(k_0) - y) + \beta V(\beta, y) : y \in [0, f(k_0)]\}.$$

From the Maximum Theorem (see the Appendix), the correspondence G is upper semi- continuous with respect to (β, k_0) (see the Appendix).

4. Multisector Optimal Growth Models With Bounded From Below Returns

In chapters 2 and 3, we studied one sector growth models. Several results were very specific to the feature of these models, for instance the monotonicity of the optimal policy which implied the convergence towards a steady state of the optimal paths. However, most optimal growth models cannot be reduced to the one-sector models similar to those of chapters 2 and 3. Let us give two examples.

Example 1

Consider a two-sector economy. At date t, sector 1 produces consumption good c_t by using a capital stock k_t^1 which is produced in sector 2. At date t, sector 2 produces capital stock k_{t+1} which will be used in period $t+1$ by the two sectors. To produce k_{t+1}, sector 2 needs a quantity k_t^2 of capital good. The social planner solves at date 0 the following problem:

$$\max \sum_{t=0}^{+\infty} \beta^t u(c_t), \beta \in]0, 1[,$$

under the constraints:

$$\forall t,\ 0 \leq c_t \leq f^c(k_t^1),$$

$$0 \leq k_{t+1} \leq f^k(k_t^2),$$

$$k_t^1 + k_t^2 \leq k_t$$

$$k_t^1 \geq 0, k_t^2 \geq 0,$$

and $k_0 \geq 0$ is given. The functions f^c and f^k are respectively the production functions of the consumption good sector and of the capital good sector.

The reader can check that, if the utility function u and the production functions f^c and f^k are strictly increasing, the initial problem becomes:

$$\max \sum_{t=0}^{+\infty} \beta^t V(k_t, k_{t+1}), 0 < \beta < 1,$$

under the constraints:

$$\forall t \geq 0, \; k_{t+1} \in \Gamma(k_t),$$

and $k_0 \geq 0$ is given. The correspondence Γ is defined by $\forall k \geq 0, \Gamma(k) = [0, f^k(k)]$, and the return function V by $V(k_t, k_{t+1}) = u(f^c(k_t - (f^k)^{-1}(k_{t+1})))$, the function $(f^k)^{-1}$ being the inverse function of f^k.

The constraint is similar to the one in chapters 2 and 3, but the utility function has another form.

Example 2 (Human capital; Lucas [50]). We have an one-sector growth model. But the output is a function of physical capital k and of effective labor N^e. Effective labor is the sum of skill-weighted manhours devoted to current production. More explicitly, assume there are N identical workers. Each worker has $h \in [0, +\infty[$ as skill level and devotes a fraction θ of his non-leisure time to current production and the remaining $(1 - \theta)$ to human capital accumulation. We thus have $N^e = Nh\theta$. Given k, h, θ, N, the level of output is $Ah^\gamma F(k, Nh\theta)$. The total productivity now is Ah^γ. The term h^γ captures the external effects of human capital while the technology level A is assumed to be constant.

We assume that the rate of growth of human capital depends, through a function G, on the non-leisure time devoted to its accumulation.

The model is as follows:

$$\max \sum_{t=0}^{+\infty} \beta^t u(c_t), 0 < \beta < 1,$$

under the constraints:

$$\forall t \geq 0, c_t + k_{t+1} - (1 - \delta)k_t \leq Ah_t^\gamma F(k_t, Nh_t\theta_t),$$

$$h_{t+1} \leq h_t(1 + G(1 - \theta_t))$$
$$h_t \geq 0, k_t \geq 0,$$

and $k_0 \geq 0, h_0 \geq 0$ are given.

Let $x = (k_0, h_0) \in \mathbb{R}^2_+$, $y = (k_1, h_1) \in \mathbb{R}^2_+$. Define the indirect utility V:

$$V(x, y) = \max_{c,\theta}\{u(c)\}$$

under the constraints:

$$c + k_1 \leq Ah_0^{\gamma} f(k_0, Nh_0\theta) + (1 - \delta)k_0,$$

$$h_1 \leq h_0(1 + G(1 - \theta)).$$

Define

$$\Phi(h_0, k_0, \theta) = Ah_0^{\gamma} f(k_0, Nh_0\theta) + (1 - \delta)k_0,$$
$$\Psi(h_0, \theta) = h_0(1 + G(1 - \theta)).$$

Define also for $x \in \mathbb{R}^2_+$:

$$\Gamma(x) = \{y \in \mathbb{R}^2_+ : \exists\theta \in [0, 1], \text{ s.t. } k_1 \leq \Phi(h_0, k_0, \theta), h_1 \leq \Psi(h_0, \theta)\}.$$

The reader can check that the initial problem is equivalent to:

$$\max \sum_{t=0}^{+\infty} \beta^t V(x_t, x_{t+1})$$

under the constraint:

$$\forall t \geq 0, x_{t+1} \in \Gamma(x_t),$$

$$x_t \in \mathbb{R}^2_+$$

and x_0 is given in $\mathbb{R}^2_+$.

In this chapter we assume that the utility is bounded below, in the sense that its absolute value is bounded by an affine function (we exclude functions such as logarithm or Cobb-Douglas functions with negative elasticities) while the technology can exhibit a non-zero maximal rate of growth (it is not the case for the models of chapters 2 and 3). We devote the next chapter to the case where the utility function could take value equal to $-\infty$ including logarithm or Cobb-Douglas functions with negative elasticities.

The plan of this chapter is as follows. In section 1, we study the general case, i.e., when the utility function may be non-concave, and the technology may have increasing returns. In section 2, we assume that the utility function is concave and the technology is convex. Section 3 will be devoted to examples.

4.1 The General Case

We consider the problem:

$$\max \sum_{t=0}^{+\infty} \beta^t F(x_t, x_{t+1})$$

under the constraints:

$$\forall t \geq 0, \ x_{t+1} \in \Gamma(x_t),$$

$$x_t \in X,$$

and x_0 is given in X, where X is a closed subset of $\mathbb{R}^n$ and contains 0.

Let us recall that an infinite sequence $(x_0, x_1, ..., x_t, ...)$ of elements in $\mathbb{R}^n$ will be denoted by $\mathbf{x}$.

A sequence $\mathbf{x}$ is *feasible* from $x_0 \in X$, if it satisfies: $\forall t \geq 0, x_{t+1} \in \Gamma(x_t)$. The set of feasible sequences from x_0 is denoted by $\Pi(x_0)$.

The assumptions are the following.

H1 The correspondence $\Gamma : X \rightarrow X$ is continuous, with non-empty, compact values. Moreover $0 \in \Gamma(0)$.

H2 There exist $\gamma \geq 0, \ \gamma' \geq 0$, such that if $y \in \Gamma(x)$ then $\|y\| \leq \gamma\|x\| + \gamma'$.

H3 The function F : graph$(\Gamma) \rightarrow \mathbb{R}$ is continuous. Moreover, there exist $A \geq 0, \ B \geq 0$ such that $\forall (x, y) \in$ graph(Γ), $|F(x, y)| \leq A + B(\|x\| + \|y\|)$.

H4 We have $\beta \in]0, 1[$ and if the constant B in assumption **H3** is strictly positive, we assume $\beta\gamma < 1$.

Remark 4.1.1. 1. The assumption $0 \in \Gamma(0)$ in **H1** is less restrictive than $\Gamma(0) = \{0\}$.
2. Assumption **H2** means that the maximal rate of growth is γ. **H2** is satisfied when $\Gamma(x) = [0, f(x)]$ where f is a positive, concave function defined on $\mathbb{R}_+$.
3. Assumption **H3** is fulfilled if the function F is concave and non-negative.
4. When the constant B in **H3** is strictly positive, condition $\beta\gamma < 1$ in **H4** ensures, as we will see below, that the sum $\sum_{t=0}^{+\infty} \beta^t F(x_t, x_{t+1})$ exists in $\mathbb{R}$.

4.1.1 Existence of an optimal solution

In order to prove the existence of solutions to the optimal growth problem, we use the product topology. We show that the function $u(\mathbf{x}) = \sum_{t=0}^{+\infty} \beta^t F(x_t, x_{t+1})$ is continuous on $\Pi(x_0)$ which is compact. A well-known result in topology allows us to conclude that an optimal solution exists (see Appendix).

Lemma 4.1.1. *(i) Under* **H2** *the following statement holds:*
$\forall x_0 \in X, \forall \mathbf{x} \in \Pi(x_0)$, *one has:*

$$\forall t \geq 1, \|x_t\| \leq \gamma^t \|x_0\| + \gamma' \sum_{j=0}^{t-1} \gamma^j.$$

(ii) Assume **H2**, **H3**. *Then there exists numbers* $c_1 \geq 0, c_2 \geq 0$ *such that:*
$\forall x_0 \in X, \forall \mathbf{x} \in \Pi(x_0)$, *one has:*

$$\forall t, |F(x_t, x_{t+1})| \leq Bc_1(\gamma^t \|x_0\| + \gamma'(1 + \gamma + \gamma^2 + \ldots + \gamma^{t-1})) + c_2,$$

where B *is the constant in assumption* **H3** *and* $c_1 > 0$.
(iii) Assume **H2**, **H3**, **H4**. *Then for every* $x_0 \in X$, *for every* $\mathbf{x} \in \Pi(x_0)$, *the sum* $\sum_{t=0}^{+\infty} \beta^t F(x_t, x_{t+1})$ *exists and is finite-valued.*

Proof. (i) For $t = 0$ the claim is obviously true. Proceed by induction to obtain the result for $t > 0$.
(ii) Using **H3** and the result obtained in (i), we have for $t \geq 0$

$$|F(x_t, x_{t+1})| \leq A+B\gamma'+B(1+\gamma)\gamma'(1+\gamma+...+\gamma^{t-1})+B(1+\gamma)\gamma^t\|x_0\|,$$

and hence, if $c_1 = \max\{(1+\gamma)\gamma', (1+\gamma)\}$ and $c_2 = A + B\gamma'$:

$$|F(x_t, x_{t+1})| \leq Bc_1(\gamma^t\|x_0\| + \gamma'(1+\gamma+\gamma^2+...+\gamma^{t-1})) + c_2.$$

Obviously, $c_1 > 0$.
(iii) Assume first $\gamma = 1$. In that case

$$\sum_{t=0}^{+\infty} \beta^t |F(x_t, x_{t+1})| \leq Bc_1 \frac{\|x_0\|}{(1-\beta)} + Bc_1\gamma' \sum_{t=0}^{+\infty} t\beta^t + c_2 < +\infty.$$

If $\gamma \neq 1$, then

$$|F(x_t, x_{t+1})| \leq Bc_1(\gamma^t\|x_0\| + \gamma'(\frac{1-\gamma^t}{1-\gamma})) + c_2,$$

and $\sum_{t=0}^{+\infty} \beta^t |F(x_t, x_{t+1})| < +\infty$, since $0 < \beta < 1$ and $\beta\gamma < 1$ when $B > 0$.

□

Lemma 4.1.2. *Assume* **H1-H4**. *Then: (i) The set $\Pi(x_0)$ is compact for the product topology.*
(ii) Let $u(\mathbf{x}) = \sum_{t=0}^{+\infty} \beta^t F(x_t, x_{t+1})$ for $\mathbf{x} \in \Pi(x_0)$. Then u is continuous.

Proof. (i) From Lemma 4.1.1(i), if $\mathbf{x} \in \Pi(x_0)$ then $\mathbf{x}$ belongs to a bounded set of the product topology. We prove that it is closed for this topology.
Indeed, let $\{\mathbf{x^n}\}$ be a sequence in $\Pi(x_0)$ which converges in the product topology to $\mathbf{x}$. Since X is closed, it is obvious that $x_t \in X, \forall t$. Since Γ is continuous, we have $x_{t+1} \in \Gamma(x_t), \forall t$. That means that $\mathbf{x} \in \Pi(x_0)$.
(ii) Let $\{\mathbf{x^n}\}$ be a sequence of $\Pi(x_0)$ which converges for the product topology to $\mathbf{x} \in \Pi(x_0)$.
From Lemma 4.1.1 (ii), if $\gamma = 1$, then, $\forall \mathbf{x} \in \Pi(x_0)$, we have

$$\forall t, |F(x_t, x_{t+1})| \leq Bc_1(\gamma^t\|x_0\| + t\gamma') + c_2,$$

and if $\gamma \neq 1$, then

$$\forall t, |F(x_t, x_{t+1})| \leq Bc_1(\gamma^t \|x_0\| + \frac{1-\gamma^t}{1-\gamma}\gamma') + c_2.$$

Hence, for any $\varepsilon > 0$, it follows from Assumption **H4** that there exists T such that for any $\mathbf{x} \in \Pi(x_0)$, for any $T' \geq T$, we have

$$\sum_{t=T'}^{+\infty} \beta^t |F(x_t, x_{t+1})| \leq \varepsilon.$$

Fix such a T'. Then

$$|u(\mathbf{x^n}) - u(\mathbf{x})| \leq \sum_{t=0}^{T'} \beta^t |F(x_t^n, x_{t+1}^n) - F(x_t, x_{t+1})| + 2\varepsilon.$$

Let n converge to $+\infty$. We get

$$\lim_{n \to +\infty} |u(\mathbf{x^n}) - u(\mathbf{x})| \leq 2\varepsilon.$$

Since ε is arbitrary, we have $\lim_{n \to +\infty} u(\mathbf{x^n}) = u(\mathbf{x})$, which proves the continuity of u. □

Proposition 4.1.1. *Assume* **H1-H4**. *Then there exists an optimal solution.*

Proof. The problem is equivalent to $\max\{u(\mathbf{x}) : \mathbf{x} \in \Pi(x_0)\}$. Let $\Pi(x_0)$ be endowed with the product topology. Then it is compact. Since, in the product topology, u is continuous, there exists a solution. □

4.1.2 Value Function and Optimal Correspondence

Let $V(x_0) = \max\{u(\mathbf{x}) : \mathbf{x} \in \Pi(x_0)\}$ be the *Value Function* of the optimal growth problem.

Proposition 4.1.2. *Assume* **H1-H4**. *Then*
(i) The function V satisfies

$$\sup_{x_0 \in X} \left\{ \frac{|V(x_0)|}{1 + B\|x_0\|} \right\} < +\infty,$$

where B is the constant in **H3***.*
(ii) The Value function V satisfies the Bellman equation:

$$\forall x_0 \in X, V(x_0) = \sup_{y \in \Gamma(x_0)} \{F(x_0, y) + \beta V(y)\}.$$

Proof. (i) From Lemma 4.1.1 (ii), there exist $A_1 > 0, A_2 \geq 0$ such that, for any $x_0 \in X$, for any $\mathbf{x} \in \Pi(x_0)$, we have

$$\sum_{t=0}^{+\infty} \beta^t |F(x_t, x_{t+1})| \leq BA_1 \|x_0\| + A_2.$$

Hence, $|V(x_0)| \leq BA_1 \|x_0\| + A_2$.
Thus

$$\sup_{x_0 \in X} \left\{ \frac{|V(x_0)|}{1 + B\|x_0\|} \right\} < +\infty.$$

(ii) Let $\mathbf{x} \in \Pi(x_0)$ satisfy $V(x_0) = \sum_{t=0}^{+\infty} \beta^t F(x_t, x_{t+1})$. Since $(x_2, x_3, ...) \in \Pi(x_1)$, by the definition of V, we have $V(x_0) \leq F(x_0, x_1) + \beta V(x_1)$, and consequently,

$$V(x_0) \leq \sup_{y \in \Gamma(x_0)} \{F(x_0, y) + \beta V(y)\}.$$

Now, let $x_1 \in \Gamma(x_0)$. There exists $(x_2, x_3, ...) \in \Pi(x_1)$ such that $V(x_1) = \sum_{t=1}^{+\infty} \beta^{t-1} F(x_t, x_{t+1})$. This implies:

$$F(x_0, x_1) + \beta V(x_1) = \sum_{t=0}^{+\infty} \beta^t F(x_t, x_{t+1}) \leq V(x_0),$$

since $(x_0, x_1, x_2, ...) \in \Pi(x_0)$. Therefore

$$\sup_{y \in \Gamma(x_0)} \{F(x_0, y) + \beta V(y)\} \leq V(x_0).$$

□

Let us now prove that V satisfies the Bellman equation with the sup replaced by max, i.e.

$$V(x_0) = \max_{y \in \Gamma(x_0)} \{F(x_0, y) + \beta V(y)\}.$$

Moreover, V is the unique solution to the Bellman equation in the space of functions h from $\mathbb{R}_+^n$ into $\mathbb{R}$ which satisfy

$$\sup_{x\in X}\left\{\frac{|h(x)|}{1+B\|x\|}\right\} < +\infty.$$

Let E denote this space. We endow it with the norm

$$\|h\| = \sup_{x\in X}\left\{\frac{|h(x)|}{1+B\|x\|}\right\}.$$

With this norm, E is a Banach space.

We introduce the following operator T:

$$\forall h \in E, \forall x \in X, Tf(x) = \sup_{y\in \Gamma(x)} \{F(x,y) + \beta h(y)\}.$$

Let us check that T maps E into E. Let $h \in E$. We have

$$\forall y \in \Gamma(x_0), \frac{|F(x_0,y)+\beta h(y)|}{1+B\|x_0\|} \leq$$

$$\frac{(A+B\gamma')+B(1+\gamma)\|x_0\|}{1+B\|x_0\|} + \beta\frac{|h(y)|}{1+B\|y\|}\frac{1+B(\gamma\|x_0\|+\gamma')}{1+B\|x_0\|}.$$

Since $\frac{|h(y)|}{1+B\|y\|} \leq \|h\|$, the second member is uniformly bounded above. We have proved that $Th \in E$.
We first show a general result in topology that if T^k (the k-iterated of T) is a contraction mapping of E for some integer k, then T has a unique fixed point.

Lemma 4.1.3. *Assume there exist an integer k and a number $\lambda \in]0,1[$ such that T^k is a λ-contraction of E, i.e.*

$$\forall h \in E, \forall g \in E, \|T^k h - T^k g\| \leq \lambda \|h-g\|.$$

Then T has a unique fixed point in E.

Proof. Obviously, T^k is a contraction. Hence, it admits a unique fixed point $v \in E$, i.e. $T^k v = v$. We claim that v is a fixed point of T.

Indeed, we have

$$\|Tv - v\| = \|T(T^k v) - T^k v\| = \|T^k(Tv) - T^k v\| \le \lambda \|Tv - v\|.$$

Thus, $Tv = v$. This fixed point is unique, since if v is a fixed point of T, then it is a fixed point of T^k. □

In the following proposition, we prove (a) that the Value function V is the unique solution in E to the Bellman equation, and, (b) as in chapter 2 and chapter 5, it is also the unique solution in the set of continuous functions (not necessary in E) which satisfy a transversality condition.

Proposition 4.1.3. *Assume* **H1-H4**. *then*
(i) The Value function V is the unique continuous solution in E to the Bellman equation:

$$\forall x_0 \in X, V(x_0) = \max_{y \in \Gamma(x_0)} \{F(x_0, y) + \beta V(y)\}.$$

(ii) We have $V = \lim_{n \to +\infty} T^n h$, for any $h \in E$.
(iii) The Value function V is the unique continuous solution to the Bellman equation, which satisfies:

$$\forall x_0 \in X, \forall \mathbf{x} \in \Pi(x_0), \lim \beta^t V(x_t) = 0.$$

(iv) We have $V = \lim_{n \to +\infty} T^n h$, for any continuous function h which satisfies the condition:

$$\forall x_0 \in X, \forall \mathbf{x} \in \Pi(x_0), \lim \beta^t h(x_t) = 0.$$

Proof. (i) It is obvious that V is a fixed point of T.
In a first step, we prove that T^k is a λ-contraction where k is an integer and $\lambda \in]0,1[$. Second, we prove that V is continuous and hence is the unique continuous fixed point of T.
We will show that there exists an integer k and a real number $\lambda \in]0,1[$ such that T^k is a λ-contraction.
Let $\varphi(x) = 1 + B\|x\|$, for $x \in X$. Take $g \in E,\ h \in E$. Since

$$\forall y \in X,\ g(y) \le h(y) + \|g - h\|\varphi(y),$$

we have

$$Tg(x_0) \leq Th(x_0) + \beta\|g - h\| \sup_{x_1 \in \Gamma(x_0)} \varphi(x_1)$$

which implies

$$Tg(x_0) \leq Th(x_0) + \beta\|g - h\|(B\gamma\|x_0\| + 1 + B\gamma').$$

By induction, we get:

$$T^k g(x_0) \leq T^k h(x_0) + \beta^k \|g - h\|(B\gamma^k\|x_0\| + 1 + B\gamma' \sum_{j=0}^{k-1} \gamma^j).$$

Reversing the role of h and g, one obtains:

$$\frac{|T^k g(x_0) - T^k h(x_0)|}{\varphi(x_0)} \leq \|g - h\|\beta^k \frac{B\gamma^k\|x_0\| + 1 + B\gamma' \sum_{j=0}^{k-1} \gamma^j}{\varphi(x_0)}.$$

If $B = 0$, then

$$\|T^k g - T^k h\| \leq \beta^k \|g - h\|.$$

In that case we can take $k = 1$ since $0 < \beta < 1$. In other words T is a β-contraction. When $B > 0$, we have

$$\frac{B\gamma^k\|x_0\| + 1 + B\gamma' \sum_{j=0}^{k-1} \gamma^j}{\varphi(x_0)} \leq \gamma^k + 1 + B\gamma' \sum_{j=0}^{k-1} \gamma^j.$$

Hence,

$$\|T^k g - T^k h\| \leq \lambda \|g - h\|,$$

with $\lambda = \beta^k \gamma^k + \beta^k (1 + \gamma' B \sum_{m=0}^{m=k-1} \gamma^m)$. For k large enough we have $0 < \lambda < 1$.

We have proved that there exists an integer k and a number $\lambda \in]0, 1[$ such that T^k is a λ-contraction. Apply Lemma 4.1.3 to conclude that V is the unique fixed point in E of T.

Since T^k is a contraction, we have $V = \lim_{n \to +\infty} (T^k)^n 0$. By the Maximum Theorem, $T0$ is continuous, and hence,$(T^k)^n 0$ is continuous for any n. Obviously, V is the uniform limit of $(T^k)^n 0$ on any compact set of X. One concludes that V is continuous. We therefore have

$$\forall x_0 \in X, V(x_0) = \max_{y \in \Gamma(x_0)} \{F(x_0, y) + \beta V(y)\}.$$

(ii) Since T^k is a contraction, the Value function V is the limit of the sequence $\{(T^k)^n h\}$ for any h in E. But we want to prove that, actually, V is the limit of the sequence $\{T^n h\}$ for any h in E. Define successively $h_1 = Th, h_2 = T^2 h, ..., h_{k-1} = T^{k-1} h$. The functions h_i belong to E. Let $\varepsilon > 0$ be given. We have:

$\exists N_i$, for $i = 0, 1, .., k-1$, such that $\forall i$, $\forall n \geq N_i$, $\|(T^k)^n h_i - V\| \leq \varepsilon$.

Let $\hat{N} = \max\{N_0, N_1, ..., N_{k-1}\}$. Let $n \geq \hat{N}k + k - 1$. Then there exists N such that $n \in \{Nk, Nk+1, ..., Nk+k-1\}$. Write $n = Nk + j$, with $j \leq k-1$. Obviously, $N \geq \hat{N}$. Therefore:

$$\|T^n h - V\| = \|(T^k)^N h_j - V\| \leq \varepsilon.$$

We have proved the statement :

$\forall \varepsilon > 0, \exists \tilde{N}$ such that, if $n \geq \tilde{N}$, then $\|T^n h - V\| \leq \varepsilon$.

In other words, we have proved that V is the limit of the sequence $\{T^n h\}$.

(iii) We first check that

$$\forall x_0 \in X, \forall \mathbf{x} \in \Pi(x_0), \lim \beta^t V(x_t) = 0.$$

Indeed, we know that that there exist $A_1 > 0, A_2 \geq 0$ such that $\forall x_0 \in X, |V(x_0)| \leq BA_1\|x_0\| + A_2$. Let $\mathbf{x} \in \Pi(x_0)$. From Lemma 4.1.1, we have

$$\forall t, |V(x_t)| \leq BA_1(\gamma^t\|x_0\| + \gamma'(1 + \gamma + ... + \gamma^{t-1})) + A_2.$$

If $B = 0$, then $\forall t, |V(x_t)| \leq A_2$ and the claim is true. If $B > 0$, since $\beta \in]0,1[, \beta\gamma < 1$, the claim is also true.

Now, let $\hat{V}$ be another solution to the Bellman equation which is continuous and satisfies the condition

$$\forall x_0 \in X, \forall \mathbf{x} \in \Pi(x_0), \lim \beta^t \hat{V}(x_t) = 0.$$

Let $\mathbf{x}$ be an optimal solution, i.e., $V(x_0) = \sum_{t=0}^{+\infty} \beta^t F(x_t, x_{t+1})$. Then

$$\hat{V}(x_0) \geq \sum_{t=0}^{T-1} \beta^t F(x_t, x_{t+1}) + \beta^T \hat{V}(x_T).$$

Let T converge to infinity. We get $\hat{V}(x_0) \geq V(x_0)$.
We prove now the converse. Since the functions $F, \hat{V}$ are continuous and the correspondence Γ is compact-valued, there exists a sequence $\mathbf{x} \in \Pi(x_0)$ such that:

$$\hat{V}(x_0) = F(x_0, x_1) + \beta\hat{V}(x_1), \hat{V}(x_1) = F(x_1, x_2) + \beta\hat{V}(x_2)...$$

and by induction:

$$\forall T, \hat{V}(x_0) = \sum_{t=0}^{T-1} \beta^t F(x_t, x_{t+1}) + \beta^T \hat{V}(x_T).$$

Take the limit when $T \to +\infty$. We obtain $\hat{V}(x_0) = u(\mathbf{x})$ and hence $\hat{V}(x_0) \leq V(x_0)$.
(iv) Let h be a continuous function which satisfies the condition

$$\forall x_0 \in X, \ \forall \mathbf{x} \in \Pi(x_0), \ \lim_{t \to +\infty} \beta^t h(x_t) = 0.$$

First observe that, by the Maximum Theorem, $T^n h$ is continuous for all n.
We will prove that $\limsup_{n \to +\infty} T^n h(x_0) \leq V(x_0)$.
Indeed, since h, F are continuous, there exists $x_1 \in \Gamma(x_0)$ such that $Th(x_0) = F(x_0, x_1) + \beta h(x_1)$. By induction, we find a sequence $\mathbf{x} \in \Pi(x_0)$ such that

$$\forall N, \ T^N h(x_0) = \sum_{t=0}^{t=N-1} \beta^t F(x_t, x_{t+1}) + \beta^N h(x_N).$$

Take the limits when $N \to +\infty$. We obtain $\limsup_N T^N h(x_0) = u(\mathbf{x})$ and hence, $\limsup_N T^N h(x_0) \leq V(x_0)$.
We now prove that $\liminf T^N h(x_0) \geq V(x_0)$. Let $\mathbf{x} \in \Pi(x_0)$. By the very definition of T, we have

$$\forall N, \ T^N h(x_0) \geq \sum_{t=0}^{t=N-1} \beta^t F(x_t, x_{t+1}) + \beta^N h(x_N).$$

Take the limits when $N \to +\infty$. We have $\liminf_N T^N h(x_0) \geq u(\mathbf{x})$. Since this inequality holds for any $\mathbf{x} \in \Pi(x_0)$, we actually have $\liminf_N T^N h(x_0) \geq V(x_0)$.
Thus, $\liminf_N T^N h(x_0) = \limsup_N T^N h(x_0) = \lim_N T^N h(x_0) = V(x_0)$. □

We now introduce the *optimal correspondence*. It is the correspondence G defined by

$$\forall x \in X, G(x) = \text{argmax}_{y \in \Gamma(x)}\{F(x,y) + \beta V(y)\}$$

or equivalently:

$$\forall x \in X, G(x) = \{y \in \Gamma(x) : V(x) = F(x,y) + \beta V(y)\}.$$

In the following proposition, we first show that the optimal policy is an upper semi-continuous correspondence. Second, we can use an algorithm which gives an approximate value of an optimal point. Hence, one can compute an approximate value of the Value function.

Proposition 4.1.4. *Assume* **H1-H4**. *Then:*
(i) G is an upper semi-continuous correspondence. If G is single-valued, then it is a continuous mapping.
(ii) Let $h \in E$. Define the correspondences G_h and G_h^k for $k = 1, 2, \ldots$ by:

$$\forall x \in X, G_h(x) = argmax_{y \in \Gamma(x)}\{F(x,y) + \beta h(y)\}$$

$$G_h^k(x) = argmax_{y \in \Gamma(x)}\{F(x,y) + \beta T^k h(y)\}.$$

Consider a sequence $\{y^k\}_{k=1,2,..}$ with $y^k \in G_h^k(x)$, $\forall k$. Then there exists a subsequence $\{y^{k_\nu}\}$ which converges to an element $y \in G(x)$ when ν converges to infinity.

Proof. (i) The statement is a consequence of the Maximum Theorem (see the Appendix).
(ii) Take $z \in \Gamma(x)$. For every k, we have:

$$F(x, y^k) + \beta T^k h(y^k) \geq F(x,z) + \beta T^k h(z).$$

From Proposition 4.1.3(ii), the function V is the limit of the sequence $\{T^k h\}$ when $k \to +\infty$. Observe also that, since $y^k \in \Gamma(x), \forall k$, there exists a subsequence $\{y^{k_\nu}\}$ which converges to some $y \in \Gamma(x)$.
Take the limits when $\nu \to +\infty$. We get:

$$F(x,y) + \beta V(y) \geq F(x,z) + \beta V(z).$$

Since z is arbitrarily chosen in $\Gamma(x)$, we conclude that $y \in G(x)$.
□

4.1.3 Properties of Optimal Paths

We can sum up these properties in the following Proposition.

Proposition 4.1.5. *Assume* **H1-H4**. *Let* $\mathbf{x} \in \Pi(x_0)$. *Then*
(i) The sequence $\mathbf{x}$ *is optimal if, and only if:*

$$\forall t \geq 0, V(x_t) = F(x_t, x_{t+1}) + \beta V(x_{t+1}).$$

(ii)The sequence $\mathbf{x}$ *is optimal if, and only if:*

$$\forall t \geq 0, x_{t+1} \in G(x_t).$$

(iii) Assume that the function F *is differentiable on* $int(\text{graph}(\Gamma))$ *(the interior of* $\text{graph}(\Gamma)$*).*
If $\mathbf{x}$ *is optimal and satisfies* $\forall t, (x_t, x_{t+1}) \in int(\text{graph}(\Gamma))$, *then* $\mathbf{x}$ *satisfies Euler equation:*

$$\forall t \geq 0, F_2(x_t, x_{t+1}) + \beta F_1(x_{t+1}, x_{t+2}) = 0$$

where F_1, F_2 *denote the derivatives of* F *with respect to the first and the second variables.*

Proof. (i) From the very definition of V, we have

$$V(x_t) \leq F(x_t, x_{t+1}) + \beta V(x_{t+1}), \forall t \geq 0.$$

Now, since $x_{t+1} \in \Gamma(x_t)$ and since V satisfies the Bellman equation, we have $V(x_t) \geq F(x_t, x_{t+1}) + \beta V(x_{t+1}), \forall t \geq 0$.
Now, let $\mathbf{x} \in \Pi(x_0)$ satisfy

$$\forall t \geq 0, V(x_t) = F(x_t, x_{t+1}) + \beta V(x_{t+1}).$$

Then, by induction, we get:

$$\forall T, V(x_0) = \sum_{t=0}^{T} \beta^t F(x_t, x_{t+1}) + \beta^{T+1} V(x_{T+1}).$$

From Proposition 4.1.3(iii), $\lim_{T \to +\infty} \beta^{T+1} V(x_{T+1}) = 0$. Hence, $V(x_0) = \sum_{t=0}^{+\infty} \beta^t F(x_t, x_{t+1})$, i.e., the sequence $\mathbf{x}$ is optimal.
We have proved statement (i).
(ii) If $\mathbf{x}$ is optimal, then by statement (i) we have $V(x_t) = F(x_t, x_{t+1}) + \beta V(x_{t+1})$. That means that $x_{t+1} \in G(x_t)$.
Conversely, if $x_{t+1} \in G(x_t)$, then $V(x_t) = F(x_t, x_{t+1}) + \beta V(x_{t+1})$. If this relation holds for every t, then by statement (i), the sequence $\mathbf{x}$ is optimal.
Since $(x_t, x_{t+1}) \in int(\text{graph}(\Gamma)), (x_{t+1}, x_{t+2}) \in int(\text{graph}(\Gamma))$, we have $(x_t, y) \in int(\text{graph}(\Gamma))$, $(y, x_{t+2}) \in int(\text{graph}(\Gamma))$ for any y in a neighborhood of x_{t+1}. The sequence $\mathbf{z}$ defined by $z_\tau = x_\tau$ for $\tau \neq t+1$, and $z_{t+1} = y$ belongs to $\Pi(x_0)$. Hence, $u(\mathbf{x}) \geq u(\mathbf{z})$. This implies:

$$F(x_t, x_{t+1}) + \beta F(x_{t+1}, x_{t+2}) \geq F(x_t, y) + \beta F(y, x_{t+2}).$$

This inequality holds for every y in a neighborhood of x_{t+1}. That means that x_{t+1} maximizes in this neighborhood the function $y \to F(x_t, y) + \beta F(y, x_{t+2})$. The result then follows. □

4.1.4 On the Continuity of the Value Function and of the Optimal Correspondence with respect to $(\boldsymbol{\beta}, \boldsymbol{x_0})$

As in Chapters 2 and 3, we write respectively $V(\beta, x_0)$, $G(\beta, x_0)$ instead of $V(x_0), G(x_0)$ for the Value function and for the optimal correspondence. We want to prove that V is continuous and G is upper semi-continuous with respect to the pair (β, x_0).
For simplicity, let us assume that the growth parameter γ of assumption **H2** equals 0. We let to the reader check that it is also true when $\gamma > 0$.
Let us go back to the proofs of Propositions 4.1.2 and 4.1.3. Consider the space E and the operator T defined in these proofs. Since

$\gamma = 0$, the operator T is a β-contraction of E. Let 0 denote the function equal to zero everywhere. Let $n > m$ and $r = n - m$. Then we have:

$$\|T^n 0 - T^m 0\| \leq \beta^m \|T^r 0\|.$$

But $\|T^r 0\| \leq \|0 - T0\| + \|T0 - T^2 0\| + ... + \|T^{r-1}0 - T^r 0\| \leq (1 + \beta + ... + \beta^{r-1})\|0 - T0\|$. Hence

$$\|T^n 0 - T^m 0\| \leq \frac{\beta^m}{1-\beta}\|T0\|.$$

Fix m and let n go to infinity. Since T^n converges to $V(\beta, .)$, we have

$$\|V(\beta, .) - T^m 0\| \leq \frac{\beta^m}{1-\beta}\|T0\|.$$

Use the definition of $\|\|$ given in the proofs of the mentioned propositions. We have

$$\sup_{x \in X} \frac{|V(\beta, x) - T^m 0(x)|}{1 + B\|x\|} \leq \frac{\beta^m}{1-\beta} \sup_{x \in X} \frac{|T0(x)|}{1 + B\|x\|}.$$

The function $T0$ is independent of β while $T^m 0$ is a continuous function of (β, x) by the Maximum Theorem. Let $\epsilon_1 \geq 0, \epsilon_2 > 0$ satisfy $0 \leq \beta - \epsilon_1, \beta + \epsilon_2 < 1$. Define $\beta_1 = \beta - \epsilon_1, \beta_2 = \beta + \epsilon_2$ and $I = [\beta_1, \beta_2]$. Then, we have

$$\sup_{x \in X} \sup_{\beta' \in I} \frac{|V(\beta', x) - T^m 0(x)|}{1 + B\|x\|} \leq \frac{\beta_2^m}{1-\beta_1} \sup_{x \in X} \frac{|T0(x)|}{1 + B\|x\|}.$$

It shows that $\frac{V(\beta', x)}{1+B\|x\|}$ is the uniform limit with respect to $(\beta', x) \in I \times X$ of the sequence of continuous functions $\{\frac{T^m 0(x)}{1+B\|x\|}\}$.
Hence $V(\beta', x)$ is continuous in $I \times X$ (see Appendix).
Since

$$G(\beta, x_0) = \text{argmax}_{y \in \Gamma(x_0)}\{F(x_0, y) + \beta V(\beta, y)\},$$

G is an upper semi-continuous correspondence by the Maximum Theorem (see Appendix).
We summarize these results in

Proposition 4.1.6. *The Value function V is continuous with respect to $(\beta, x) \in [0, 1[\times X$. The optimal correspondence G is upper semi-continuous with respect to $(\beta, x) \in I \times X$.*

4.2 The Case with Concave Return Function and Convex Technology

We assume that $X \neq \{0\}$ and $\Gamma(X) \neq \{0\}$ (if not, the optimal solution is immediate: it is the null sequence $(0, 0, ...0, ...)$. We replace **H2** by **H'2**:
H'2: $graph(\Gamma)$ is convex (i.e., if $y_1 \in \Gamma(x_1), y_2 \in \Gamma(x_2), \lambda \in [0, 1]$, then $\lambda y_1 + (1 - \lambda)y_2 \in \Gamma(\lambda x_1 + (1 - \lambda)x_2)$).
The following Lemma shows that **H1** and **H'2** imply **H2**.

Lemma 4.2.1. *Assume* **H1** *and* **H' 2**. *Then there exist* $\gamma' > 0$ *and* $\gamma \geq 0$ *such that:* $y \in \Gamma(x) \Longrightarrow \|y\| \leq \gamma' + \gamma\|x\|$.

Proof. Let $a \in X$, $a \neq 0$ and $\Gamma(a) \neq \{0\}$. Let $y \in \Gamma(x)$ with $\|x\| \geq \|a\|$. Since $(0, 0) \in graph(\Gamma)$, by **H'2** we have $\frac{\|a\|}{\|x\|}y \in \Gamma(\frac{\|a\|x}{\|x\|})$. By **H1** one can define

$$\hat{\gamma} = \max_{z \in \Gamma(X \cap S_a)} \{\|z\|\}$$

where $S_a = \{x \in \mathbb{R}^n : \|x\| = \|a\|\}$. Hence, if $\|x\| \geq \|a\|$ and if $y \in \Gamma(x)$ then $\|y\| \leq \frac{\hat{\gamma}}{\|a\|}\|x\|$. Define $\gamma = \frac{\hat{\gamma}}{\|a\|}$.
Now let $\gamma' = \max\{\|z\| : z \in \Gamma(x), \|x\| \leq \|a\|\}$. By **H1**, such a γ' exists. Moreover, $\gamma' \geq \max\{\|z\| : z \in \Gamma(a)\} > 0$.
Summing up, if $y \in \Gamma(x)$, then $\|y\| \leq \gamma' + \gamma\|x\|$. □

We now replace **H3** by **H'3**.
H'3: F is a concave, non negative function from $graph(\Gamma)$ into $\mathbb{R}$.
The following Lemma shows that actually, **H'3** implies **H3**.

Lemma 4.2.2. *If* **H'3** *holds, then* **H3** *also holds.*

Proof. For sake of simplicity, we assume that $int(graph(\Gamma))$ is nonempty. Since F is concave, it has a subdifferential everywhere in $int(graph(\Gamma))$ (see Appendix). Let $(\hat{x}, \hat{y}) \in int(graph(\Gamma))$ and (p, q) be in the subdifferential of F at $(\hat{x}, \hat{y})$. We then have:

$$\forall (x, y) \in graph(\Gamma), F(\hat{x}, \hat{y}) - F(x, y) \geq p \cdot (\hat{x} - x) + q \cdot (\hat{y} - y),$$

and hence,

$$0 \leq F(x, y) \leq F(\hat{x}, \hat{y}) - p \cdot \hat{x} - q \cdot \hat{y} + p \cdot x + q \cdot y$$

$$\leq |F(\hat{x},\hat{y})| + \|p\|\|\hat{x}\| + \|q\|\|\hat{y}\| + B(\|x\| + \|y\|),$$

where $B = \max\{\|p\|, \|q\|\}$. Take $A = |F(\hat{x},\hat{y})| + \|p\|\|\hat{x}\| + \|q\|\|\hat{y}\|$ to end the proof. □

All the results of the previous section hold for this section. But we have more.

Proposition 4.2.1. *Assume* **H1, H'2, H'3,** *and* **H4**. *Then*
(i) The Value function V is concave and non-negative.
(ii) It is the unique solution to the Bellman equation which is concave and non-negative.

Proof. (i) Let $\mathbf{x}$ be feasible from x_0 and $\mathbf{x}'$ be feasible from x'_0 with $x_0 \neq x'_0$. Let $\lambda \in [0,1]$. By **H'2**,

$$\forall t, \lambda x_{t+1} + (1-\lambda)x'_{t+1} \in \Gamma(\lambda x_t + (1-\lambda)x'_t).$$

Hence

$$V(\lambda x_0 + (1-\lambda)x'_0) \geq \sum_{t=0}^{+\infty} \beta^t F(\lambda x_t + (1-\lambda)x'_t, \lambda x_{t+1} + (1-\lambda)x'_{t+1})$$

$$= \sum_{t=0}^{+\infty} \beta^t F(\lambda(x_t, x_{t+1}) + (1-\lambda)(x'_t, x'_{t+1}))$$

$$\geq \sum_{t=0}^{+\infty} \beta^t [\lambda F(x_t, x_{t+1}) + (1-\lambda)F(x'_t, x'_{t+1})] \text{ (by } \mathbf{H'3})$$

$$= \lambda \sum_{t=0}^{+\infty} \beta^t F(x_t, x_{t+1}) + (1-\lambda) \sum_{t=0}^{+\infty} \beta^t F(x'_t, x'_{t+1}).$$

Since the above inequality holds for all the feasible paths starting from x_0 and x'_0, it holds also for the optimal paths. Thus

$$V(\lambda x_0 + (1-\lambda)x'_0) \geq \lambda V(x_0) + (1-\lambda)V(x'_0).$$

It is obvious that V is non-negative since F is non-negative.
(ii) Consider the operator T associated with the Bellman equation. We have shown that there exists an integer k such that T^k is a

contraction mapping of the space E of functions h from X into $\mathbb{R}$ which satisfy $\sup_{x\in X} \frac{|h(x)|}{1+B\|x\|} < +\infty$. Moreover, V is a fixed point of T in E. Since any concave, non-negative function h on X belongs to E because it satisfies $\sup_{x\in X} \frac{|h(x)|}{1+B\|x\|} < +\infty$ (see Appendix), we conclude that V is the unique solution to the Bellman equation which is concave, non-negative on X. □

The following proposition gives conditions under which a feasible path from x_0 is optimal. Observe that one of the conditions is that X is a subset of the positive orthant $\mathbb{R}^n_+$.

Proposition 4.2.2 (Mangasarian Lemma).
Assume **H1**, **H'2**, **H'3**, **H4**, $X \subset \mathbb{R}^n_+$, *X contains* 0, *F is differentiable in* $int(graph(\Gamma))$ *and* $F_2 \leq 0$.
If $\mathbf{x}$ *is a feasible path from* x_0 *which satisfies* $\forall t, (x_t, x_{t+1}) \in int(graph(\Gamma))$ *and*

$$(i) \textit{ Euler equation: } \forall t, F_2(x_t, x_{t+1}) + \beta F_1(x_{t+1}, x_{t+2}) = 0$$

and

$$(ii) \textit{ Transversality Condition: } \lim_{t\to\infty} \beta^t F_1(x_t, x_{t+1}) \cdot x_t = 0$$

then $\mathbf{x}$ *is optimal.*

Proof. Let $\mathbf{x}$ be feasible from x_0 and satisfy the conditions of the statement of the Proposition. Let $\mathbf{x}' \in \Pi(x_0)$. By the concavity of F, we have:

$$\Delta_T = \sum_{t=0}^{t=T} \beta^t F(x_t, x_{t+1}) - \sum_{t=0}^{t=T} \beta^t F(x'_t, x'_{t+1})$$

$$\geq \sum_{t=0}^{t=T} \beta^t [F_1(x_t, x_{t+1}) \cdot (x_t - x'_t) + F_2(x_t, x_{t+1}) \cdot (x_{t+1} - x'_{t+1})]$$

$$= \sum_{t=0}^{t=T-1} \beta^t (F_2(x_t, x_{t+1}) + \beta F_1(x_{t+1}, x_{t+2})) \cdot (x_{t+1} - x'_{t+1})$$

$$+\beta^T F_2(x_T, x_{T+1}) \cdot (x_{T+1} - x'_{T+1}).$$

Since the Euler equation holds and $x'_{T+1} \in \mathbb{R}^n_+$ and $F_2(x_T, x_{T+1}) \leq 0$, we have

$$\Delta_T \geq \beta^T F_2(x_T, x_{T+1}) \cdot x_{T+1} = -\beta^{T+1} F_1(x_{T+1}, x_{T+2}) \cdot x_{T+1}.$$

Hence $\lim_{T \to +\infty} \Delta_T \geq 0$ and $\mathbf{x}$ is optimal. □

Let us now give sufficient conditions which ensure that the Value function is differentiable.

Proposition 4.2.3 (Benveniste-Scheinkman).
Assume **H1**, **H'2**, **H'3**, **H4** *and that F is differentiable on the interior of graph(Γ).*
Let (x, y) satisfy $y \in G(x)$ and $(x, y) \in int(graph(\Gamma))$. Then the derivative of V at x, $V'(x)$, exists and equals $F_1(x, y)$.

Proof. First, we have $V(x) = F(x, y) + \beta V(y)$. Second, since $(x, y) \in int(graph(\Gamma))$, there exists a neighborhood $\mathcal{U}(x)$ of x such that $y \in \Gamma(x'), \forall x' \in \mathcal{U}(x)$.
Define for $x' \in \mathcal{U}(x)$, $\phi(x') = F(x', y) + \beta V(y)$. Then ϕ is concave. We have:

$$\forall x' \in \mathcal{U}(x), V(x') \geq \phi(x').$$

Since x belongs to the interior of X, the subdifferential of V at x is non-empty. Let p be in this subdifferential. We have:

$$\forall x' \in \mathcal{U}(x), V(x) - V(x') \geq p \cdot (x - x').$$

Since $V(x) = \phi(x)$ and $V(x') \geq \phi(x')$, the above inequality implies

$$F(x, y) + \beta V(y) - (F(x', y) + \beta V(y)) \geq p \cdot (x - x'),$$

or equivalently,

$$F(x, y) - F(x', y) \geq p \cdot (x - x').$$

This inequality holds for any $x' \in \mathcal{U}(x)$. Hence $p = F_1(x, y)$. This also shows that the subdifferential of V at x is a singleton. One concludes that V is differentiable at x (see Appendix). Obviously we have $V'(x) = F_1(x, y)$. □

Remark 4.2.1. One can remark that the proof of the differentiability of V is fairly simple. The proof that the optimal policy is differentiable and the Value function V is twice differentiable turns out to be much harder and requires more assumptions on the utility function F (see e.g. Santos [74], Montrucchio [61]).

4.2.1 The One Dimension Case

In the one-sector models of chapters 2 and 3, the optimal paths always converge to a steady state. In the introduction, we have given an example of a two-sector model that can be transformed into a one dimensional optimal growth model. However the optimal path may be non-monotonic.

Proposition 4.2.4. *Assume* $X = \mathbb{R}_+$ *and* **H1**, **H'2**, **H'3**, **H4**. *Assume also that the function* F *is strictly concave with respect to the second variable. We then have:*
(i) The optimal policy is a continuous function g.
(ii) Assume furthermore: F *is twice continuously differentiable in* $int(graph(\Gamma))$ *and* $\forall x \in \mathbb{R}_+, (x, g(x)) \in int(graph(\Gamma))$. *Then*
(a) If the cross derivative F_{12} *is positive (respectively negative) in* $int(graph(\Gamma))$, *then* g *is increasing (respectively decreasing).*
(b) If moreover, $\gamma < 1$ *in assumption* **H2** *then, if* $F_{12} > 0$, *any optimal path converges to a steady state* x^* *(i.e.* $x^* = g(x^*)$*), and if* $F_{12} < 0, (x, g(x)) \in int(graph\Gamma), \forall x$, *any optimal sequence converges either to a steady state or to a two-period cycle.*

Proof. (i) We have $g(x) = \text{argmax}\{F(x,y) + \beta V(y) : y \in \Gamma(x)\}$. Since $F(x, .)$ is strictly concave, the optimal solution is unique. The optimal policy is a function g which is continuous by the Maximum Theorem.
(ii) Since $\forall x, (x, g(x)) \in int(graph\Gamma)$, from Proposition 4.2.3, we obtain that V is differentiable and $V'(x) = F_1(x, g(x))$. Since $g(x)$ is an interior solution, we have $F_2(x, g(x)) + \beta V'(g(x)) = 0, \forall x \in \mathbb{R}_+$. Assume $F_{12} > 0$ and that there exist $x < x'$, with $g(x) \geq g(x')$. Then:

$$0 = F_2(x', g(x')) + \beta V'(g(x')) > F_2(x, g(x')) + \beta V'(g(x'))$$

$$\geq F_2(x, g(x)) + \beta V'(g(x)) = 0$$

(the second inequality follows from the concavity of F and V): a contradiction. Hence $x < x' \Rightarrow g(x) < g(x')$.
The same argument applies when $F_{12} < 0$ to claim that $x < x' \Rightarrow g(x) > g(x')$.

Now assume that $\gamma < 1$ in assumption **H2**. From Lemma 4.1.1 (i) we get that:

$$\forall \mathbf{x} \in \Pi(x_0), \forall t, 0 \leq x_t \leq \gamma^t x_0 + \frac{\gamma'}{1-\gamma} \leq x_0 + \frac{\gamma'}{1-\gamma}.$$

Hence, any optimal path satisfies

$$\forall t, x_t \in [0, \hat{A}] \text{ with } \hat{A} = x_0 + \frac{\gamma'}{1-\gamma}.$$

When $F_{12} > 0$, the optimal sequence $\mathbf{x}$ is monotonic (either decreasing, or increasing), bounded. It converges to a steady state, since g is continuous and since $x_{t+1} = g(x_t), \forall t$.
Consider the case where $F_{12} < 0$. Let $\mathbf{x}$ be an optimal sequence from x_0.
If $x_0 = x_1$, then $x_t = x_0, \forall t$, and the claim is obviously true.
Assume, without loss of generality, that $x_0 < x_1$. Then $x_2 = g(x_1) < x_1 = g(x_0)$ (g is decreasing).
First, assume $x_2 \leq x_0$. Then $x_3 = g(x_2) \geq g(x_0) = x_1$, and so on. By induction, we have:

$$x_{2n} \leq x_{2(n-1)} \leq ... \leq x_0 < x_1 \leq x_3 \leq ... \leq x_{2n+1}.$$

Since $x_t \in [0, \hat{A}], \forall t$, the sequence $\{x_{2n}\}$ converges to $\underline{x}$ and $\{x_{2n+1}\}$ converges to $\overline{x}$, with $\overline{x} > \underline{x}$.
Since $x_{2n+1} = g(x_{2n})$, $x_{2n} = g(x_{2n-1}), \forall n$, we have therefore $\overline{x} = g(\underline{x})$ and $\underline{x} = g(\overline{x})$. This implies $g^2(\overline{x}) = \overline{x}$, $g^2(\underline{x}) = \underline{x}$. The optimal sequence converges to a two-periods cycle.
Now assume $x_0 < x_2$. Then $x_3 = g(x_2) < g(x_0) = x_1$. But, since $x_2 < x_1$, we have $x_3 = g(x_2) > g(x_1) = x_2$. By induction:

$$x_0 < x_2 < x_4 < ...x_{2n} < x_{2n+1} < x_{2n-1} < ... < x_1.$$

The increasing sequence $\{x_{2n}\}_{n=0,1,...}$ converges to $\overline{x}$ and the decreasing sequence $\{x_{2n+1}\}_{n=0,1,...}$ converges to $\underline{x}$. If $\overline{x} = \underline{x}$, then the optimal sequence converges to a steady state. If $\overline{x} < \underline{x}$, the optimal sequence converges to a two-periods cycle, because $g^2(\underline{x}) = \underline{x}$ and $g^2(\overline{x}) = \overline{x}$. □

Remark 4.2.2. We let the reader prove the following:
Assume that F is strictly concave with respect to the second variable, twice continuously differentiable on $int(graph(\Gamma))$.
If $F_{12} \geq 0$ (respectively ≤ 0), then g is non decreasing (respectively non increasing), i.e. $x > x'$ (respectively $x < x'$) $\Rightarrow g(x) \geq g(x')$ (respectively $g(x) \leq g(x')$).

4.3 Examples

4.3.1 Example 1 (The Ramsey Model)

We go back to the Ramsey Model:

$$\max \sum_{t=0}^{+\infty} \beta^t u(c_t), \text{ with } 0 < \beta < 1,$$

under the constraints

$$\forall t, c_t + k_{t+1} \leq f(k_t),$$

$$c_t \geq 0, k_t \geq 0, k_0 > 0 \text{ is given.}$$

We assume that the function u is strictly concave, strictly increasing, differentiable, $u(0) = 0$, that the function f is concave, strictly increasing, differentiable, $f(0) = 0, f'(+\infty) < 1$.
We know that the problem is equivalent to

$$\max \sum_{t=0}^{+\infty} \beta^t F(k_t, k_{t+1}),$$

$$\text{with } 0 < \beta < 1, \ F(k, y) = u(f(k) - y),$$

under the constraints

$$\forall t, 0 \leq k_{t+1} \leq f(k_t), \ k_0 > 0 \text{ is given.}$$

We have proved in chapter 2 that, under the assumptions on f, there exists a unique point $\bar{k}$ such that $\bar{k} = f(\bar{k}), k > \bar{k} \Rightarrow f(k) <$

$k, k < \bar{k} \Rightarrow f(k) > k$. Let $\hat{A} > \max\{\bar{k}, k_0\}$. Then, if $\mathbf{k} \in \Pi(k_0)$, we have $\forall t, k_t \leq \hat{A}$.
This model satisfies the assumptions **H1, H2, H3, H4**.
First, we can set $X = [0, \hat{A}]$.
In **H1**, the correspondence Γ is defined by $\Gamma(k) = [0, f(k)]$. Obviously, Γ maps X into X and is continuous.
In **H2**, we can take $\gamma = 0, \gamma' = \hat{A}$.
Since F is concave, non negative, it satisfies **H3**.
Obviously, **H4** is satisfied.
Hence there exists an optimal solution. The value function is the unique continuous, concave solution to the Bellman equation, by Proposition 4.2.1. From Proposition 4.2.4, the optimal path is monotonic and converges to a steady state. Recall that if $u'(0) = +\infty$, then the optimal consumptions are strictly positive at each period.

4.3.2 Example 2

This example is a Ramsey model with a linear utility function. It does not satisfy the Inada condition. We will show that, when the initial capital stock is low, the optimal consumptions will be equal to zero up to some date T_0 and equal the steady state consumption after T_0. In other words, it is optimal to consume nothing during the first T_0 periods.

Consider the problem:

$$\max \sum_{t=0}^{+\infty} \beta^t c_t, \text{ with } 0 < \beta < 1,$$

under the constraints

$$\forall t, c_t + k_{t+1} \leq f(k_t),$$

$$c_t \geq 0, k_t \geq 0, k_0 > 0 \text{ is given.}$$

We assume that the function f is strictly concave, strictly increasing, $f(0) = 0$, $f'(0) > \frac{1}{\beta}$, $f'(+\infty) < 1$.
This model is equivalent to

$$\max \sum_{t=0}^{+\infty} \beta^t F(k_t, k_{t+1}),$$

$$\text{with } 0 < \beta < 1,\ F(k,y) = f(k) - y,$$

under the constraints

$$\forall t, 0 \leq k_{t+1} \leq f(k_t),\ k_0 > 0 \text{ is given.}$$

As before, this model satisfies the assumptions of Proposition 4.2.1 and the remark which follows Proposition 4.2.2. Hence there exists an optimal path which is monotonic and which converges to a steady state.
More precisely, let k^s satisfies $f'(k^s) = \frac{1}{\beta}$.
We claim that if $f(k_0) \geq k^s$, then the sequence $\mathbf{k} \in \Pi(k_0)$ defined by $k_t = k^s, \forall t \geq 1$ is optimal.
Proof of the claim First assume that $f(k_0) > k^s$. The sequence $\mathbf{k}$ defined above (i) is feasible, (ii) interior, (iii) satisfies Euler equation which is $-1 + \beta f'(k_{t+1}) = 0$, and (iv) satisfies the transversality condition. From Proposition 4.2.2, this sequence is optimal.
Now assume $f(k_0) = k^s$. Let g denote the optimal policy. For any $k_0' > k_0$, from the proof given above, we have $g(k_0') = k^s$. Since g is continuous (Proposition 4.1.4), by letting k_0' converge to k_0, we obtain $k^s = g(k_0)$. Hence, the sequence $\mathbf{k}$ defined by $k_t = k^s, \forall t \geq 1$ is optimal from k_0.
We now claim that if $f(k_0) < k^s$, then the sequence $\mathbf{k} \in \Pi(k_0)$ defined by $k_t = f^t(k_0)$ for $t = 0, ..., T$, and $k_t = k^s$, for $t > T$, where T is the first period such that $f^{T+1}(k_0) \geq k^s$, is optimal from k_0.
Proof of the claim Take another feasible path $\mathbf{k}' \in \Pi(k_0)$. Take some integer $N > T$. Let

$$\Delta_N = \sum_{t=0}^{N} \beta^t F(k_t, k_{t+1}) - \sum_{t=0}^{N} \beta^t F(k_t', k_{t+1}').$$

By concavity of F, we get

$$\Delta_N \geq (\beta f'(k_1) - 1)(k_1 - k_1') + ... + \beta^{T-1}(\beta f'(k_T) - 1)(k_T - k_T')$$

$$+\beta^T(\beta f'(k^s)-1)(k^s-k'_{T+1})+...$$
$$+\beta^{N-1}(\beta f'(k^s)-1)(k^s-k_N)-\beta^N(k^s-k_{N+1}).$$

Since, by the definition of T, we have $k_t = f^t(k_0) < k^s, \forall t \leq T$ and hence, $f'(k_t) > f'(k^s) = \frac{1}{\beta}, \forall t \leq T$. We also have $\forall t \leq T, k'_t \leq f^t(k_0) = k_t$. Thus:

$$\Delta_N \geq -\beta^N k^s.$$

Therefore $\lim_{N\to+\infty} \Delta_N \geq 0$. This ends the proof of the claim. Observe that the optimal consumptions is equal to zero up to T.

4.3.3 Example 3: A Consumption-Savings Problem

The following model is in Stokey and Lucas [65]. Consider a consumer who has the intertemporal utility

$$\sum_{t=0}^{t=+\infty} \beta^t u(c_t)$$

where $c_t \geq 0, \forall t$ and the utility u is strictly concave, strictly increasing, $u(0) = 0, u'(0) = +\infty, u'(+\infty) = 0$.
The consumer has a strictly positive income I at each period, and an initial wealth $x_0 \geq 0$ in period 0. The consumer cannot borrow. The interest rate r is positive and constant over time. The interest factor $R = (1+r)$ satisfies $1 < R < \frac{1}{\beta}$. At period $t+1$, the consumer's wealth is $x_{t+1} = R(x_t + I - c_t)$, where c_t is his consumption at period t. The problem of the consumer is:

$$\max \sum_{t=0}^{t=+\infty} \beta^t u(c_t)$$

under the constraints

$$\forall t \geq 0, 0 \leq c_t \leq I + x_t - \frac{x_{t+1}}{R},$$

$x_0 \geq 0$ is given.
Since $\beta R < 1$ and $I > 0$, we will show that the consumer will exhaust his initial wealth x_0 in finite time and consumes his income I in every period thereafter.

The model is equivalent to:

$$\max \sum_{t=0}^{+\infty} \beta^t F(x_t, x_{t+1})$$

under the constraints:

$$\forall t \geq 0, 0 \leq x_{t+1} \leq R(I + x_t),$$

$x_0 \geq 0$ is given. Here $F(x, y) = u(I + x - y/R)$ and $X = \mathbb{R}_+$. The reader can check that the assumptions **H1**, **H'2**, **H'3**, **H4** are satisfied. There exists an optimal solution. Since F is strictly concave with respect to the second variable, the optimal policy is a continuous function g (Proposition 4.2.4). Moreover, the cross derivative F_{12} is positive, and hence, the optimal path will be monotonic.
Let $m > 0$ satisfy $u'(I + m) = R\beta u'(I)$. Since $u'(+\infty) = 0$ and since u' is decreasing, such an m exists.
Claim 1: $x_0 \in [0, m] \Rightarrow g(x_0) = 0$.
Proof of the claim Take a feasible sequence $\mathbf{x} \in \Pi(x_0)$, i.e., a sequence which satisfies $0 \leq x_{t+1} \leq R(x_t + I), \forall t \geq 0$. Compute the difference:

$$\Delta_T = u(I + x_0) + \beta u(I) + ... + \beta^T u(I)$$

$$-[u(I + x_0 - \frac{x_1}{R}) + \beta u(I + x_1 - \frac{x_2}{R}) + ... + \beta^T u(I + x_T - \frac{x_{T+1}}{R})].$$

We find:

$$\Delta_T \geq u'(I + x_0)\frac{x_1}{R} - \beta u'(I)x_1 + \beta u'(I)x_2(\frac{1}{R} - \beta) +$$

$$... + \beta^{T-1}u'(I)x_T(\frac{1}{R} - \beta) + \beta^T u'(I)\frac{x_{T+1}}{R}.$$

Since $u'(I + x_0) \geq u'(I + m) = R\beta u'(I)$ and $\frac{1}{R} > \beta$, we have:

$$\Delta_T \geq \beta^T u'(I)\frac{x_{T+1}}{R} \geq 0,$$

and hence the sequence $(x_0, 0, 0, ...0, ...)$ is optimal.
Claim 2: If $x_0 > m$, then $0 < g(x_0) < (I + x_0)R$.

Proof of the claim From the Inada condition ($u'(0) = +\infty$), the optimal consumption must be strictly positive, i.e., $(I + x_0)R - g(x_0) > 0$. Let us prove that $g(x_0) > 0$. Suppose on the contrary, $x_1 = g(x_0) = 0$. The optimal consumption sequence from x_0 is $c_0 = I + x_0, c_t = I, \forall t \geq 1$. Let us define another feasible consumption sequence $(c'_0, c'_1, ..., c'_t, ...)$:
$c'_0 = I + x_0 - \frac{\varepsilon}{R}, c'_1 = I + \varepsilon, c'_t = I, \forall t \geq 2$, where $\varepsilon > 0$ is chosen such that $I + x_0 - \frac{\varepsilon}{R} \geq 0$.
Compare the utilities given by these sequences of consumptions. We have

$$\Delta_\varepsilon = \sum_{t=0}^{+\infty} \beta^t u(c'_t) - \sum_{t=0}^{+\infty} \beta^t u(c_t) =$$

$$u(I + x_0 - \frac{\varepsilon}{R}) + \beta u(I + \varepsilon) - u(I + x_0) - \beta u(I)$$

$$\geq u'(I + x_0 - \frac{\varepsilon}{R})(-\frac{\varepsilon}{R}) + \beta u'(I + \varepsilon)\varepsilon$$

$$= \varepsilon[\beta u'(I + \varepsilon) - \frac{1}{R}u'(I + x_0 - \frac{\varepsilon}{R})].$$

When $\varepsilon \to 0$, the expression $[\beta u'(I + \varepsilon) - \frac{1}{R}u'(I + x_0 - \frac{\varepsilon}{R})] \to \beta u'(I) - \frac{1}{R}u'(I + x_0) > \beta u'(I) - \frac{1}{R}u'(I + m) = 0$.
Hence $\Delta_\varepsilon > 0$ for $\varepsilon > 0$ small enough. That contradicts the optimality of the sequence **c**. This ends the proof of claim 2.
Claim 3: If $x_0 > m$, then $x_1 = g(x_0) < x_0$.
Proof of the claim If $x_1 \leq m$, the claim is true. So, assume $x_1 > m$. From Claim 2, $x_2 = g(x_1)$ satisfies $0 < x_2 < R(I + x_1)$. From Proposition 4.2.3, we have $V'(x_1) = u'(I + x_1 - \frac{x_2}{R})$. Since we also have $0 < x_1 < (I+x_0)R$, by the same proposition one gets $V'(x_0) = u'(I + x_0 - \frac{x_1}{R})$. We also have the Euler equation:

$$\frac{1}{R}u'(I + x_0 - \frac{x_1}{R}) = \beta u'(I + x_1 - \frac{x_2}{R}),$$

or equivalently

$$V'(x_0) = R\beta V'(x_1).$$

Since $R\beta < 1$, we therefore have $x_1 < x_0$. Claim 3 is proved.
Claim 4: Let $x_0 \geq 0$ be given. Then there exists T such that $x_T \leq m$. Hence $x_t = 0, \forall t > T$.

Proof of the claim If $x_0 \leq m$, then the proof follows from Claim 1. Hence, assume $x_0 > m$ and by contradiction that the sequence $\mathbf{x}$ satisfies $x_t > m, \forall t$. From Claim 3, this sequence is strictly decreasing. It converges to $x^* \geq m$. Since the optimal policy is continuous, we must have $x^* = g(x^*)$. But if $x^* > m$, from Claim 3, we must have $g(x^*) < x^*$. Hence $x^* = m$. But Claim 1 implies that $g(x^*) = 0$. In both cases, we obtain a contradiction with $x^* = g(x^*)$. Hence, there exists T with $x_T \leq m$. From claim 1, $x_t = 0, \forall t > T$.

4.3.4 Example 4: A Two-Sector Model

We now present a two-sector optimal growth model which can be transformed in an one-dimensional growth model. What we want to show is that the optimal policy of this model is not monotonic.

In our economy, there are two goods: (i) a consumption good c produced in sector 1 through a production function $f^c(k^1, l^1)$ using a capital stock k^1 and a quantity of labor l^1, and (ii) a capital good k produced in sector 2 through a production function f^k using k^2 capital and l^2 labor. We assume that the supply of labor is constant over time and equal to 1. The functions f^c and f^k are concave, strictly increasing and satisfy $f^c(0,l) = f^c(k,0) = f^k(0,l) = f^k(k,0) = 0, \forall k \geq 0, \forall l \geq 0$.

The consumer solves the problem:

$$\max \sum_{t=0}^{+\infty} \beta^t u(c_t), \text{ with } 0 < \beta < 1,$$

under the constraints:

$$\forall t \geq 0, \; 0 \leq c_t \leq f^c(k_t^1, l_t^1),$$
$$0 \leq k_{t+1} \leq f^k(k_t^2, l_t^2),$$
$$k_t^1 + k_t^2 \leq k_t,$$
$$l_t^1 + l_t^2 \leq 1,$$
$$l_t^1 \geq 0, \; l_t^2 \geq 0, \; k_t^1 \geq 0, \; k_t^2 \geq 0,$$

and $k_0 > 0$ is given.

Introduce the indirect utility F defined as follows

$$F(k,y) = \max_{k^1,l^1,k^2,l^2} f^c(k^1,l^1),$$

under the constraints

$$0 \le k^1 + k^2 \le k,$$

$$0 \le l^1 + l^2 \le 1,$$
$$0 \le y \le f^k(k^2,l^2),$$
$$k^1 \ge 0,\ k^2 \ge 0,\ l^1 \ge 0,\ l^2 \ge 0.$$

We assume that the utility function is $u(c) = c$. We let to the reader to prove that the initial problem is equivalent to the following one:

$$\max \sum_{t=0}^{+\infty} \beta^t F(k_t, k_{t+1}),$$

under the constraints

$$0 \le k_{t+1} \le f^k(k_t, 1),\ \forall t \ge 0,$$

and k_0 is given.
We suppose that $f^c(k,l) = k^{1-\alpha} l^\alpha$, with $1 > \alpha > 0$, and that $f^k(k,l) = \min\{l, \frac{k}{\lambda}\}$, with $1 > \lambda > 0$.
One can check that

$$F(k,y) = (k - \lambda y)^{1-\alpha}(1-y)^\alpha.$$

The problem becomes

$$\max \sum_{t=0}^{+\infty} \beta^t (k_t - \lambda k_{t+1})^{1-\alpha}(1-k_{t+1})^\alpha,$$

$$\text{such that } \forall t \ge 0,\ 0 \le k_{t+1} \le \min\{1, \frac{k_t}{\lambda}\},$$

and $k_0 > 0$ is given.
Here we can take $X = \mathbb{R}_+$.

In **H1**, the correspondence Γ is defined by $\Gamma(x) = [0, \min\{1, \frac{x}{\lambda}\}]$. Obviously $0 \in \Gamma(0)$.
In **H2**, we take $\gamma = 0$ and $\gamma' = 1$, since $y \in \Gamma(x) \Rightarrow 0 \leq y \leq 1$.
The function F being concave, nonnegative, assumption **H3** is satisfied.
H4 is satisfied since $0 < \beta < 1$ and $\gamma = 0$.
All the results of the first section apply. In particular, the Value function is the unique solution to the Bellman Equation, which is concave, continuous, and the optimal policy is continuous.
But the cross derivative of F is

$$F_{12}(k, y) = \alpha(1 - \alpha)(1 - y)^{\alpha-1}(k - \lambda)^{-\alpha-1}(\lambda - k).$$

Then $F_{12} > 0$ if $k < \lambda$ and $F_{12} < 0$ if $k > \lambda$. The optimal policy is increasing if $k < \lambda$ and decreasing if $k > \lambda$.

4.3.5 Example 5: A Human Capital Model

We present a simplified version of the Lucas model which is given in Stokey and Lucas [65], p.111. In this version, there is no physical capital and no externality.
Assume that at date t, the growth rate of capital is given by the formula

$$\theta_t = \phi(\frac{h_{t+1}}{h_t}),$$

where h_t is the human capital at date t and θ_t is the number of working hours. We assume that $0 \leq \theta_t \leq 1$.
We assume that ϕ is continuous, decreasing and satisfies $\phi(1+\lambda) = 0,\ \phi(1-\delta) = 1$, where $\lambda > 0,\ 0 \leq \delta < 1$. In other words, we assume that without training ($\theta_t = 1$) the human capital depreciates with rate δ and if the worker devotes his whole time for training, his human capital will grow at rate λ. Hence, λ is the maximal rate of growth of human capital.

The consumption good is produced through a production function using only effective labor. At date t, effective labor is $\theta_t h_t N_t$ with N_t denoting the number of workers at date t. We assume that $N_t = 1,\ \forall t$.

The model is

$$\max \sum_{t=0}^{+\infty} \beta^t u(c_t),$$

$$\text{such that } \forall t,\ 0 \le c_t \le f(\theta_t h_t),$$

$$\theta_t = \phi(\frac{h_{t+1}}{h_t}),\ 0 \le \theta_t \le 1,$$

and $h_0 > 0$ is given.

We make the following assumptions:
(i) $u(c) = c^\mu,\ 0 < \mu < 1$,
(ii) $\beta > 0$,
(iii) $f(L) = L^\alpha,\ 0 < \alpha < 1$,
(iv) $\beta(1+\lambda) < 1$.
It is easy to check that this model is equivalent to:

$$\max \sum_{t=0}^{+\infty} \beta^t [h_t \phi(\frac{h_{t+1}}{h_t})]^{\alpha\mu},$$

under the constraints

$$\forall t \ge 0,\ (1-\delta)h_t \le h_{t+1} \le (1+\lambda)h_t,$$

and $h_0 > 0$ is given.

Let $X = \mathbb{R}_+$ and Γ be defined by $\Gamma(x) = [(1-\delta)x,\ (1+\lambda)x]$. For $(x,y) \in graph(\Gamma)$, define the function F by $F(x,y) = [x\phi(\frac{y}{x})]^{\alpha\mu}$, if $x > 0$, and $F(0,y) = 0$ for $y \ge 0$. Since $0 \le \phi(\frac{x}{y}) \le 1$, $0 \le F(x,y) \le x^{\alpha\mu}$. Thus F is continuous on $graph(\Gamma)$.

The model is obviously of the form:

$$\max \sum_{t=0}^{+\infty} \beta^t F(h_t, h_{t+1})$$

$$\text{such that } \forall t \ge 0, h_{t+1} \in \Gamma(h_t),$$

and $h_0 > 0$ is given.
Assumptions **H1-H4** are satisfied. Hence there exists an optimal solution.
The Value function V satisfies the Bellman Equation :

$$\forall h \geq 0,\ V(h) = \max_{y\in[(1-\delta)h,\ (1+\lambda)h]}\{F(h,y) + \beta V(y)\}.$$

We claim that (i) $V(h) = Ah^{\alpha\mu}$ for some constant A, and (ii) there exists $u^* \in [1-\delta,\ 1+\lambda]$ such that the optimal path $\mathbf{h}$ from h_0 is $h_t = (u^*)^t h_0,\ \forall t \geq 0$.

Proof of the claim Let T denote the operator which associates with any continuous function f on $\mathbb{R}_+$ the function $Tf(h) = \max_{y\in[(1-\delta)h,\ (1+\lambda)h]}\{F(h,y) + \beta f(y)\}$. From Proposition 4.1.3, we know that $V = \lim_{n\to+\infty} T^n 0$. Take $h > 0$. We have successively:

$$\begin{aligned} T0(h) &= \max_{y\in[(1-\delta)h,\ (1+\lambda)h]}\{(h\phi(\frac{y}{h}))^{\alpha\mu}\} \\ &= h^{\alpha\mu} \max_{u\in[1-\delta,\ 1+\lambda]}\{\phi(u)^{\alpha\mu}\} = A_1 h^{\alpha\mu}, \\ T^2 0(h) &= \max_{y\in[(1-\delta)h,\ (1+\lambda)h]}\{(h\phi(\frac{y}{h}))^{\alpha\mu} + \beta A_1 y^{\alpha\mu}\} \\ &= h^{\alpha\mu} \max_{u\in[1-\delta,\ 1+\lambda]}\{\phi(u)^{\alpha\mu} + \beta A_1 u^{\alpha\mu}\} = A_2 h^{\alpha\mu}. \end{aligned}$$

By induction, we have $T^n 0(h) = A_n h^{\alpha\mu}$. Hence $A_n \to A$ when $n \to +\infty$. In other words, $V(h) = Ah^{\alpha\mu}$. We have proved the first part of the claim.

The Value function satisfies the Bellman Equation:

$$\begin{aligned} \forall h > 0,\ V(h) &= \max_{y\in[(1-\delta)h,\ (1+\lambda)h]}\{F(h,y) + \beta A y^{\alpha\mu})\} \\ &= h^{\alpha\mu} \max_{u\in[1-\delta,\ 1+\lambda]}\{\phi(u)^{\alpha\mu} + \beta A u^{\alpha\mu}\} \\ &= h^{\alpha\mu}\{\phi(u^*)^{\alpha\mu} + \beta A (u^*)^{\alpha\mu}\}, \end{aligned}$$

where $u^* \in [1-\delta,\ 1+\lambda]$. Therefore, the optimal policy g is defined by $g(h) = u^* h$ and $\mathbf{h}$ is optimal if, and only if,$h_t = (u^*)^t h_0,\ \forall t \geq 0$. We have proved the second part of the claim.

4.3.6 Example 5: Learning by Doing Model

This example is given in Stokey and Lucas [65] p. 108. Consider a monopolist producing a new product; his production function displays learning by doing. We suppose that the cost function C depends on the production at date t, q_t, and on the cumulative production Q_t, i.e., $Q_t = Q_{t-1} + q_{t-1}$. More precisely, the production cost C_t

is $C_t = C(q_t, Q_t)$. We assume that C is convex, continuously differentiable and satisfies $\forall Q \geq 0,\ C(0, Q) = 0,\ 0 < \underline{c} \leq \frac{\partial C}{\partial q}(0, Q) < \overline{c}$. We also assume that given q, the unit-cost function $\frac{C(q,Q)}{q}$ is a decreasing function with respect to Q.
The price is given by an inverse demand function $\psi : \mathbb{R}_+ \to \mathbb{R}_+$ which is continuously differentiable, strictly decreasing, and such that the income function $q\psi(q)$ is strictly concave. We assume that $\psi(0) > \overline{c}$ and $\psi(+\infty) < \underline{c}$.
The monopolist maximizes his intertemporal profit

$$\max \sum_{t=0}^{+\infty} \beta^t [(Q_{t+1} - Q_t)\psi(Q_{t+1} - Q_t) - C(Q_{t+1} - Q_t, Q_t)],$$

$$\text{with } 0 < \beta < 1,$$

under the constraints

$$\forall t \geq 0, Q_{t+1} \geq Q_t,$$

and $Q_0 \geq 0$ is given.
Define the return function F:

$$F(x, y) = (y - x)\psi(y - x) - C(y - x, x), \text{ for } y \geq x.$$

The model becomes

$$\max \sum_{t=0}^{+\infty} \beta^t F(x_t, x_{t+1}),$$

such that

$$x_{t+1} \geq x_t \geq 0,\ \forall t \geq 0,$$

and x_0 is given.
A priori, this problem does not fall in the category of problems we study in this chapter since the technology correspondence Γ is not compact valued ($\Gamma(x) = [x,\ +\infty[$). We will show that actually we can restrict $\Gamma(x)$ to a compact set.

Let V denote the Value function. First, observe that for any $x_0 \geq 0$, the stationary sequence $(x_0, x_0, ..., x_0, ...)$ is feasible. Thus

$V(x_0) \geq 0$. Second, observe that $F(x,y) \leq (y-x)\psi(y-x) - \underline{c}(y-x)$ since, by convexity of C and the fact that $C(0,y) = 0$, we have $C(x,y) = C(x,y) - C(0,y) \geq \frac{\partial C}{\partial q}(0,y)x \geq \underline{c}x$.
Consider the function f defined by $f(x) = x\psi(x) - \underline{c}x$. This function is strictly concave, satisfies $f(0) = 0$. Moreover $\frac{f(x)}{x}$ converges to $\psi(0) - \underline{c} > \overline{c} - \underline{c} > 0$ when x converges to zero, which implies $f'(0) > 0$. We also have that $\frac{f(x)}{x}$ converges to $\psi(+\infty) - \underline{c} < 0$ when x converges to $+\infty$. This shows that $f(x)$ converges to $-\infty$ when x converges to $+\infty$. Since, by concavity, $f'(x) \leq \frac{f(x)}{x}$, we have $f'(+\infty) < 0$. Thus, there exists a unique maximum point $\bar{x}$ of f. Let $M = f(\bar{x})$.
Since $F(x,y) \leq f(y-x)$, we have

$$\forall x_0 \geq 0,\ 0 \leq V(x_0) \leq \frac{M}{1-\beta}.$$

We also have

$$\sum_{t=0}^{+\infty} \beta^t F(x_t, x_{t+1})$$

$$= \sum_{t=0}^{+\infty} \beta^t [(x_{t+1} - x_t)\psi(x_{t+1} - x_t) - C(x_{t+1} - x_t, x_t)]$$

$$\leq (x_1 - x_0)\psi(x_1 - x_0) - C(x_1 - x_0, x_0) + \frac{\beta M}{1-\beta}$$

$$\leq (x_1 - x_0)\psi(x_1 - x_0) - \underline{c}(x_1 - x_0) + \frac{\beta M}{1-\beta} = f(x_1 - x_0) + \frac{\beta M}{1-\beta}.$$

The function $g(x) = f(x) + \frac{\beta M}{1-\beta}$ is strictly concave, satisfies $g(0) = \frac{\beta M}{1-\beta} > 0,\ g'(0) > 0, g(+\infty) = -\infty$. Therefore, there exists a unique point $\hat{x}$ such that $g(x) \geq 0 \Leftrightarrow x \leq \hat{x}$. Since $V(x_0) \geq 0$, we must choose x_1 such that $x_1 - x_0 \leq \hat{x}$.

The problem now becomes

$$\max \sum_{t=0}^{+\infty} \beta^t F(x_t, x_{t+1})$$

under the constraints

$$\forall t \geq 0,\ 0 \leq x_t \leq x_{t+1} \leq x_t + \hat{x},$$

and $x_0 \geq 0$ is given.

We first check that the function F satisfies **H3**. Indeed, we here have $\Gamma(x) = [x\ , x+\hat{x}]$. Let $(x, y) \in graph(\Gamma)$. Since, by assumption, the function $q\psi(q)$ is concave, we have $F(x, y) \leq (y - x)\psi(y - x) - \underline{c}(y - x) \leq B(y - x) + A$ for some constants A and B. Thus, $F(x, y) \leq |B|(|x| + |y|) + |A|$.
Now, we have $F(x, y) \geq -C(y-x, y)$. Since $C(y-x, y) \leq C(\hat{x}, y) \leq C(\hat{x}, 0)$, one gets $F(x, y) \geq -C(\hat{x}, 0)$.
Summing up,

$$|F(x, y)| \leq |B|(|x| + |y|) + |A| + C(\hat{x}, 0).$$

We let the reader check that this model satisfies the remaining assumptions **H1, H2, H4**.

5. Optimal Growth Models With Unbounded From Below Returns

Models of economic growth are often used by macroeconomists working in the area of real business cycles. Often in these applications utility functions are assumed to be logarithmic or homogeneous of degree θ where θ may be negative. While these assumptions greatly simplify the analysis, they introduce problems of unboundedness from below at the origin. As a consequence, well-established results about the continuity of the Value function and about the upper semi-continuity of the optimal policy may not hold. In this chapter, for simplicity, we will consider a one sector model. The instantaneous utility function has finite values for non-null consumption and equals $-\infty$ if the consumption equals zero.
Alvarez and Stokey [3] assume, for multisector models, that the return functions are homogeneous of degree θ or logarithmic . They also assume that the technology exhibits constant returns to scale. Le Van and Morhaim [47] do not impose constant returns to scale for the technology. They prove, as Alvarez and Stokey, that the optimal solutions exists, that the Value function may be continuous and give algorithms to compute the Value function. For models with recursive utility functions which are unbounded from below, the reader can refer to Boyd [18] or Le Van and Vailakis [48].
The chapter is organized as follows. We present the model and its assumptions in section 1. Existence of the solutions will be proved in section 2. The properties of the Value function are given in section 3. Section 4 exhibits some properties of the optimal paths. Examples are given in section 5.

5.1 The Model

Consider the model

$$\max \sum_{t=0}^{+\infty} \beta^t u(f(k_t) - k_{t+1})$$

under the constraints

$$\forall t \geq 0, g(k_t) \leq k_{t+1} \leq f(k_t),$$

where $k_0 \geq 0$ is given.
We make the following assumptions:
H1 The function f is continuous, $f(0) = 0, f(k) \geq 0, \forall k > 0$. Moreover, this function is continuously differentiable.
The function g is continuous, $g \leq f,\ g(k) \geq 0, \forall k \geq 0$.
There exist $\gamma \geq 0, \gamma' \geq 0$, such that: $g(k) \leq y \leq f(k) \Rightarrow y \leq \gamma k + \gamma'$.
H2 The function u is strictly concave, increasing, continuously differentiable, $u(0) = -\infty$.
H3 $0 < \beta < 1;\ \beta\gamma < 1$.
We denote by $\Pi(k_0)$ the set of sequences of capital stocks $\mathbf{k}$, feasible from k_0) , i.e., the set of $\mathbf{k}$ which satisfy

$$\forall t,\ g(k_t) \leq k_{t+1} \leq f(k_t).$$

Let us introduce the following assumption:
H4 For every $k_0 > 0$, there exists a sequence $\mathbf{k} \in \Pi(k_0)$ such that $\sum_{t=0}^{+\infty} \beta^t u(f(k_t) - k_{t+1}) > -\infty$.
Let $\Pi'(k_0) = \{\mathbf{k} \in \Pi(k_0) : \sum_{t=0}^{+\infty} \beta^t u(f(k_t) - k_{t+1}) > -\infty\}$.
Observe that, as in Chapters 2, 3 and 4, if we introduce the function U defined on $\Pi(k_0)$ by $U(\mathbf{k}) = \sum_{t=0}^{+\infty} \beta^t u(f(k_t) - k_{t+1})$, then the initial problem has a compact form:

$$\max\{U(\mathbf{k}) : \mathbf{k} \in \Pi(k_0)\}.$$

Comment The reader may wonder why we use the constraints

$$\forall t,\ g(k_t) \leq k_{t+1} \leq f(k_t).$$

Obviously, if the function g equals zero, then we have the standard Ramsey model. But if we impose the investment to be irreversible, that is, non negative at every period, (see e.g. Mitra [57], Mitra and Ray [59], Aliprantis, Border and Burkinshaw [2], Duran and Le Van [36]), then the constraints in the Ramsey model become

$$\forall t, \ (1-\delta)k_t \leq k_{t+1} \leq f(k_t),$$

where $f(k) = F(k) + (1-\delta)k$ (F is the production function) and δ is the capital depreciation rate.

5.2 Existence of Optimal Solutions

To prove existence of optimal solutions, we proceed in two steps. As in Chapters 2, 3, and 4, we prove that $\Pi(k_0)$ is compact for the product topology. Second, and this differs from the previous chapters, we prove that the function U is upper semi-continuous for the same topology (i.e. if $\{\mathbf{k^n}\}$ is a sequence of elements in $\Pi(k_0)$ which converge to $\mathbf{k} \in \Pi(k_0)$, then $\limsup U(\mathbf{k^n}) \leq U(\mathbf{k})$). We then apply the well-known result stating that to maximize an upper semi-continuous function over a compact set always has a solution.

Lemma 5.2.1. *Assume* **H1** *and* **H3**. *If* $\mathbf{k} \in \Pi(k_0)$, *then:*

$$\forall t, \ 0 \leq k_t \leq \gamma^t k_0 + \gamma'(1+\gamma+ ... +\gamma^{t-1}).$$

Proof. See Chapter 4. □

The proof of the following lemma is left to the reader.

Lemma 5.2.2. *Assume* **H1** *and* **H3**. *The set* $\Pi(k_0)$ *is compact for the product topology.*

Lemma 5.2.3. *Assume* **H1, H2** *and* **H3**. *Then the function* U *is upper semi-continuous, for the product topology, on* $\Pi(k_0)$.

Proof. First, observe that, since u is concave, there exist $A \geq 0, B \geq 0$ such that $u(c) \leq Ac + B, \forall c \geq 0$. Thus, for every $\mathbf{k} \in \Pi(k_0)$, for every $t \geq 0$, we have

$$u(f(k_t) - k_{t+1}) \leq u(f(k_t)) \leq Af(k_t) + B \leq A(\gamma k_t + \gamma') + B.$$

From Lemma 5.2.1, we have

$$u(f(k_t) - k_{t+1}) \leq A\gamma' + B + A\gamma(\gamma^t k_0 + \gamma'(1 + \gamma + ... + \gamma^{t-1})).$$

Second, by **H3**, we can say that given $\varepsilon > 0$, there exists T_0 such that, for any $T \geq T_0$, for any $\mathbf{k} \in \Pi(k_0)$, we have

$$\sum_{t=T}^{+\infty} \beta^t u^+(f(k_t) - k_{t+1}) \leq \varepsilon.$$

Here, u^+ denotes the positive value of u.
Now, let $\mathbf{k^n} \in \Pi(k_0)$ converge to $\mathbf{k} \in \Pi(k_0)$. Let $\varepsilon > 0$ be given. Then there exists T_0 such that, for any n, for any $T \geq T_0$, we have

$$U(\mathbf{k^n}) \leq \sum_{t=0}^{T} \beta^t u(f(k_t^n) - k_{t+1}^n) + \sum_{t=T+1}^{+\infty} \beta^t u^+(f(k_t^n) - k_{t+1}^n)$$

$$\leq \sum_{t=0}^{T} \beta^t u(f(k_t^n) - k_{t+1}^n) + \varepsilon.$$

Therefore

$$\limsup U(\mathbf{k^n}) \leq \sum_{t=0}^{T} \beta^t u(f(k_t) - k_{t+1}) + \varepsilon.$$

Let T converge to infinity. We get

$$\limsup U(\mathbf{k^n}) \leq U(\mathbf{k}) + \varepsilon.$$

Since ε is arbitrarily chosen, we actually have $\limsup U(\mathbf{k^n}) \leq U(\mathbf{k})$. We have proved that U is upper semi-continuous. □

Proposition 5.2.1. *Assume* **H1, H2** *and* **H3**. *Then there exists an optimal solution.*
If we add **H4**, *then the maximal value is not equal to* $-\infty$.

Proof. Existence of optimal solutions follows from Lemma 5.2.2 and Lemma 5.2.3. The second statement is trivial. □

5.3 Value Function

As in the previous chapters, the Value function is defined by

$$V(k_0) = \max \sum_{t=0}^{+\infty} \beta^t u(f(k_t) - k_{t+1})$$

under the constraints:

$$\forall t \geq 0, \ g(k_t) \leq k_{t+1} \leq f(k_t),$$

and $k_0 \geq 0$ is given.

Proposition 5.3.1. *Assume* **H1, H2** *and* **H3**. *The Value function is upper semi-continuous and satisfies:*
(i) $V(0) = -\infty$
(ii) There exist $A_1 \geq 0, B_1 \geq 0$ *such that:*

$$\forall k_0 \geq 0, V(k_0) \leq A_1 k_0 + B_1.$$

(iii) $\forall k_0 \geq 0, \ \forall \mathbf{k} \in \Pi(k_0), \limsup \beta^t V(k_t) \leq 0.$
Add assumption **H4**. *Then*
For any $k_0 > 0, V(k_0) > -\infty$ *and for any* $\mathbf{k} \in \Pi'(k_0)$, *we have* $\lim_{t \to +\infty} \beta^t V(k_t) = 0$.

Proof. We will prove that V is upper semi-continuous. Let $\{k_0^n\}$ be a sequence converging to k_0. Let $\{\mathbf{k_0^n}\}$ be a sequence of elements in $\Pi(k_0^n)$. From the proof of Lemma 5.2.3, there exist $A \geq 0, B \geq 0$ such that:

$$\forall t, u(f(k_t^n) - k_{t+1}^n) \leq A\gamma' + B + A\gamma(\gamma^t k_0^n + \gamma'(1 + \gamma + ... + \gamma^{t-1})).$$

Let $\{k_0^{n_k}\}$ satisfy $\lim_k V(k_0^{n_k}) = \limsup_n V(k_0^n)$. Let $\{\mathbf{k^{n_k}}\} \subset \Pi(k_0^{n_k})$ be such that $V(k_0^{n_k}) = \sum_{t=0}^{+\infty} \beta^t u(f(k_t^{n_k}) - k_{t+1}^{n_k})$. Let $\varepsilon > 0$ be given. Then there exist N, T_0 such that

$$\forall k \geq N, \forall T \geq T_0, \sum_{t=T+1}^{+\infty} \beta^t u^+(f(k_t^{n_k}) - k_{t+1}^{n_k}) \leq \varepsilon.$$

Hence,

$$V(k_0^{n_k}) \leq \sum_{t=0}^{T} \beta^t u(f(k_t^{n_k}) - k_{t+1}^{n_k}) + \varepsilon. \quad (1)$$

From Lemma 5.2.1, the sequence $\{\mathbf{k}^{\mathbf{n_k}}\}$ belongs to a fixed compact set, for the product topology, when k is large enough. We can assume that this sequence converges to some sequence $\mathbf{k}$. Since we have

$$\forall t, \forall k, g(k_t^{n_k}) \leq k_{t+1}^{n_k} \leq f(k_t^{n_k}),$$

we obtain:

$$\forall t, \ g(k_t) \leq k_{t+1} \leq f(k_t).$$

In other words, $\mathbf{k} \in \Pi(k_0)$.
Taking the limits in inequality (1), we have:

$$\lim_k V(k_0^{n_k}) \leq \sum_{t=0}^{T} \beta^t u(f(k_t) - k_{t+1}) + \varepsilon.$$

Let T converge to infinity. Then

$$\lim_k V(k_0^{n_k}) \leq U(\mathbf{k}) + \varepsilon \leq V(k_0) + \varepsilon.$$

Hence, $\lim_k V(k_0^{n_k}) \leq V(k_0)$. We have proved the upper semi-continuity of V.
It is obvious that $V(0) = -\infty$ since $u(0) = -\infty$.
To prove statement (ii), we go back to the proof of Lemma 5.2.1. For any $\mathbf{k} \in \Pi(k_0)$, for any t, we have

$$u(f(k_t) - k_{t+1}) \leq A\gamma' + B + A\gamma(\gamma^t k_0 + \gamma'(1 + \gamma + ... + \gamma^{t-1})).$$

By **H3**, we have $\sum_{t=0}^{+\infty} \beta^t u(f(k_t) - k_{t+1}) \leq A_1 k_0 + B_1$ for some $A_1 \geq 0, B_1 \geq 0$. Therefore, $V(k_0) \leq A_1 k_0 + B_1$. Regarding (iii), we apply Lemma 5.2.1. Then:

$$V(k_t) \leq A_2 \gamma^t k_0 + B_2,$$

for some $A_2 \geq 0, B_2 \geq 0$. Again, by assumption **H3**, we obtain $\limsup_t \beta^t V(k_t) \leq 0$.
It is obvious that if we add **H4**, then $V(k_0) > -\infty$ for any $k_0 > 0$.
Let $\mathbf{k} \in \Pi(k_0)$. Then we have:

$$-\infty < U(\mathbf{k}) \leq \sum_{t=0}^{T} \beta^t u(f(k_t) - k_{t+1}) + \beta^{T+1} V(k_{T+1}).$$

Take the limits. We obtain

$$0 = \lim_{T \to +\infty} [U(\mathbf{k}) - \sum_{t=0}^{T} \beta^t u(f(k_t) - k_{t+1})] \leq \liminf_{T} \beta^{T+1} V(k_{T+1}).$$

Since $\limsup \beta^t V(k_t) \leq 0$, we have therefore $\lim \beta^t V(k_t) = 0$. □

Let S be the set of functions h from $\mathbb{R}_+$ into $\mathbb{R} \cup \{-\infty\}$ which are upper semi-continuous and satisfy
(i) for any $k_0 \geq 0$, for any sequence $\mathbf{k} \in \Pi(k_0)$, we have

$$\limsup_{t} \beta^t h(k_t) \leq 0,$$

(ii) for any $k_0 > 0$, for any sequence $\mathbf{k} \in \Pi'(k_0)$, then $\lim_t \beta^t h(k_t) = 0$.

Proposition 5.3.2 (Bellman Equation). *Assume* **H1, H2** *and* **H3**. *Then:*
The Value function satisfies the Bellman Equation (B)

$$\forall k_0 \geq 0, V(k_0) = \max\{u(f(k_0) - y) + \beta V(y) : g(k_0) \leq y \leq f(k_0)\}.$$

(ii) The Value function is the unique solution in S to the Bellman equation.

Proof. The proof of statement (i) is left to the reader. We now prove (ii). Let $\hat{V}$ be another solution in S to the Bellman equation. Since $\hat{V}$ is upper semi-continuous, there exists a sequence $\mathbf{k} \in \Pi(k_0)$ such that

$$\hat{V}(k_0) = \sum_{t=0}^{T} \beta^t u(f(k_t) - k_{t+1}) + \beta^{T+1} \hat{V}(k_{T+1}).$$

Take the limits when T goes to infinity. We have

$$\hat{V}(k_0) \leq U(\mathbf{k}) + \limsup_{T} \beta^{T+1} \hat{V}(k_{T+1}) \leq U(\mathbf{k}) \leq V(k_0).$$

If $k_0 = 0$, then $V(k_0) = -\infty$, and $\hat{V}(k_0) = -\infty$.
If $k_0 > 0$ and if $\mathbf{k} \in \Pi'(k_0)$, then, from the very definition of $\hat{V}$:

$$\hat{V}(k_0) \geq \sum_{t=0}^{T} \beta^t u(f(k_t) - k_{t+1}) + \beta^{T+1}\hat{V}(k_{T+1}),$$

and hence, by letting T converge to infinity,

$$\hat{V}(k_0) \geq \sum_{t=0}^{+\infty} \beta^t u(f(k_t) - k_{t+1}) + \lim_T \beta^{T+1}\hat{V}(k_{T+1}) \geq U(\mathbf{k}).$$

Since $\mathbf{k}$ is arbitrary in $\Pi'(k_0)$, we actually have $\hat{V}(k_0) \geq V(k_0)$. So, we have proved that $\hat{V} = V$ and the uniqueness of the solution in S. □

We say that a function $\varphi : \mathbb{R}_+ \to \mathbb{R} \cup \{-\infty\}$ is continuous in the generalized sense if it is continuous at any strictly positive point, and if k converges to zero, then $\varphi(k)$ converges to $-\infty$.

Proposition 5.3.3 (Continuity of the Value function). *Assume* **H1, H2, H3, H4** *and add the following assumption:*

There exists a continuous (in the generalized sense) function φ such that $V(k_0) \geq \varphi(k_0), \forall k_0 > 0$, and, if $\mathbf{k} \in \Pi'(k_0)$, then $\lim_t \beta^t \varphi(k_t) = 0$.
Under these assumptions, the Value function V is continuous in the generalized sense.

Proof. If $\{k_0^n\}$ is a sequence converging to zero, then, since V is upper semi-continuous, we have $\limsup_n V(k_0^n) \leq V(0) = -\infty$. Thus $V(k_0^n) \to -\infty$.
Now, assume $k_0 > 0$ and let $\{k_0^n\}$ be a sequence converging to k_0. Let $\mathbf{k} \in \Pi(k_0)$ be an optimal solution. We claim that there exists a sequence $\{\mathbf{k^n}\} \in \Pi(k_0^n), \forall n$ and $\mathbf{k^n} \to \mathbf{k}$ when $n \to +\infty$.
Assume first $g(k_0) < k_1 < f(k_0)$. Then for n large enough, we have $g(k_0^n) < k_1 < f(k_0^n)$. The sequence defined for n large enough by $k_t^n = k_t, \forall t \geq 1$ belongs to $\Pi(k_0^n)$ and trivially converges to $\mathbf{k}$.
Now assume $k_1 = f(k_0)$. If, for any t, we have either $k_t = f(k_{t-1})$ or $k_t = g(k_{t-1})$, then define $\mathbf{k^n}$ by $k_t^n = f(k_{t-1}^n)$ if $k_t = f(k_{t-1})$, and $k_t^n = g(k_{t-1}^n)$ if $k_t = g(k_{t-1})$. This sequence converges to $\mathbf{k}$.

Assume that there exists T, the first date such that $g(k_{T-1}) < k_T < f(k_{T-1})$. Then for n large enough, say $n \geq N$, we will have $g(k^n_{T-1}) < k_T < f(k^n_{T-1})$. The sequence defined by

$$\text{(a) } k^n_t = f(k^n_{t-1}) \text{ if } k_t = f(k_{t-1}) \text{ and if } t < T$$

$$\text{(b) } k^n_t = g(k^n_{t-1}) \text{ if } k_t = g(k_{t-1}) \text{ and if } t < T$$

$$\text{(c) } k^n_t = k_t \text{ if } t \geq T$$

belongs to $\Pi(k^n_0)$ and converges to $\mathbf{k}$.
If $k_1 = g(k_0)$, the same argument applies. Our claim is true.
Let N be given. Then, one has:

$$V(k^n_0) \geq \sum_{t=0}^{N} \beta^t u(f(k^n_t) - k^n_{t+1}) + \beta^{N+1} V(k^n_{N+1}).$$

Since $V(k_0) > -\infty$, we have $k_t > 0, \forall t$. Hence $k^n_{N+1} > 0$ for n large enough. Therefore

$$V(k^n_0) \geq \sum_{t=0}^{N} \beta^t u(f(k^n_t) - k^n_{t+1}) + \beta^{N+1} \varphi(k^n_{N+1}).$$

Let $n \to +\infty$. We obtain:

$$\liminf_n V(k^n_0) \geq \sum_{t=0}^{N} \beta^t u(f(k_t) - k_{t+1}) + \beta^{N+1} \varphi(k_{N+1}).$$

Let $N \to +\infty$. Then $\liminf_n V(k^n_0) \geq V(k_0)$.
Since, from Proposition 5.3.1,

$$\limsup_n V(k^n_0) \leq V(k_0),$$

we have

$$\lim_n V(k^n_0) = V(k_0).$$

This ends the proof of the continuity of V. □

Proposition 5.3.4 (Differentiability of the Value Function). *Assume* **H1, H2, H3, H4**.
(i) Assume moreover that f is increasing and $g = 0$. Then the Value function V has left and right derivatives at any point $k_0 > 0$. Moreover, V is differentiable at k_0 iff the optimal path from k_0 is unique.
The Value function V is differentiable almost everywhere in $\mathbb{R}_+$, or, equivalently, there exists a unique optimal path from almost everywhere in $\mathbb{R}_+$.
(ii) Assume that g is convex, and f is concave, increasing. In that case, the Value function V is concave.
Let $\mathbf{k}^*$ be the optimal path from k_0. If we have $g(k_0) < k_1^* < f(k_0)$ (it is the case when $g(k_0) = 0$), then the Value function is differentiable at k_0 and

$$V'(k_0) = u'(f(k_0) - k_1^*)f'(k_0).$$

Proof. (i) The proof is similar to the one of Chapter 3.
(ii) The proof of the concavity of V is left to the reader.
Since $V(k_0) > -\infty, V(0) = -\infty$, and $u(0) = -\infty, f(0) = 0$, then, when $g(k_0) = 0$, we must have $k_1^* > 0$. The proof of the differentiability of V is similar to the one in Chapter 2. □

Let T be the mapping associated with the Bellman equation. To any upper semi-continuous function $h : \mathbb{R}_+ \to \mathbb{R} \cup \{-\infty\}$, let us associate the function Th:

$$\forall x \in \mathbb{R}_+,\ Th(x) = \max\{u(f(x) - y) + \beta h(y) : g(x) \leq y \leq f(x)\}.$$

The function Th is upper semi-continuous since u, f, and g are continuous, and h is upper semi-continuous (by the Maximum Theorem, see Appendix).

Proposition 5.3.5 (Algorithms to compute the Value function). *Assume* **H1, H2, H3, H4**. *We have:*

$$\text{(i) } \forall k_0 \geq 0, V(k_0) = \lim_{n \to +\infty} T^n a(k_0)$$

where a is a constant number.

$$\textit{(ii)}\ \forall k_0 \geq 0, V(k_0) = \lim_{n\to+\infty} T^n h(k_0),$$

where h belongs to S and satisfies $Th \leq h$.

$$\textit{(iii)}\ \forall k_0 \geq 0, V(k_0) = \lim_{n\to+\infty} T^n h(k_0),$$

where h is upper semi-continuous and satisfies:
(a) $Th \leq h$,
(b) $\forall k_0 \geq 0, \forall \mathbf{k} \in \Pi(k_0), \limsup_t \beta^t h(k_t) \leq 0$,
(c) $\forall k_0 \geq 0, \forall \mathbf{k} \in \Pi(k_0), U(\mathbf{k}) \leq h(k_0)$.

Proof. (i) We first prove that $\liminf_n T^n a(k_0) \geq V(k_0)$. Indeed, let $\mathbf{k} \in \Pi(k_0)$. We have

$$Ta(k_0) \geq u(f(k_0) - k_1) + \beta a$$

$$T^2 a(k_0) \geq u(f(k_0) - k_1) + \beta T a(k_1)$$

$$\geq u(f(k_0) - k_1) + \beta u(f(k_1) - k_2) + \beta^2 a,$$

and so on...
By induction:

$$T^n a(k_0) \geq \sum_{t=0}^{n-1} \beta^t u(f(k_t) - k_{t-1}) + \beta^n a,$$

and hence $\liminf_n T^n a(k_0) \geq U(\mathbf{k})$.
This inequality holds for any $\mathbf{k} \in \Pi(k_0)$. Thus $\liminf_n T^n a(k_0) \geq V(k_0)$.
We now prove that $\limsup_n T^n a(k_0) \leq V(k_0)$.
For any n, there exist $k_1^n, k_2^n, ..., k_n^n$ with $g(k_t^n) \leq k_{t+1}^n \leq f(k_t^n)$ for $t = 0, 1, ..., n-1$, and

$$T^n a(k_0) = u(f(k_0) - k_1^n) + \beta u(f(k_1^n) - k_2^n) + ...$$

$$+\beta^{n-1} u(f(k_{n-1}^n) - k_n^n) + \beta^n a.$$

The sequence $(k_0, k_1^n, ..., k_n^n, g(k_n^n), g(k_n^n), ...)$ belongs to $\Pi(k_0)$. One may assume that it converges to some $\mathbf{k} = (k_0, k_1, ..., k_n, k_{n+1}, ...) \in \Pi(k_0)$. From the proof of Lemma 5.2.3, we see that given $\varepsilon > 0$, there exists N_0 such that for any $N \geq N_0$, for any $n > N$, we have

$$\beta^N u(f(k_N^n) - k_{N+1}^n) + ... + \beta^{n-1} u((f(k_{n-1}^n) - k_n^n) \leq \varepsilon.$$

Choose such an N. Then for any $n > N$, we have:

$$T^n a(k_0) \leq u(f(k_0) - k_1^n) + \beta u(f(k_1^n) - k_2^n) + ...$$
$$+ \beta^{N-1} u(f(k_{N-1}^n) - k_N^n) + \varepsilon + \beta^n a.$$

Fix N and let n converge to infinity. Then:

$$\limsup_n T^n a(k_0) \leq u(f(k_0) - k_1) + \beta u(f(k_1) - k_2) + ...$$
$$+ \beta^{N-1} u(f(k_{N-1}) - k_N) + \varepsilon.$$

Now let N go to infinity. Then

$$\limsup_n T^n a(k_0) \leq U(\mathbf{k}) + \varepsilon \leq V(k_0) + \varepsilon,$$

and hence

$$\limsup_n T^n a(k_0) \leq V(k_0).$$

(ii) Let $k_0 > 0$ and let $\mathbf{k}$ satisfy $V(k_0) = \sum_{t=0}^{+\infty} \beta^t u(f(k_t) - k_{t+1})$. By **H4**, the sequence $\mathbf{k}$ must belong to $\Pi'(k_0)$.We obtain by induction:

$$\forall n, T^n h(k_0) \geq u(f(k_0) - k_1) + \beta u(f(k_1) - k_2) + ...$$
$$+ \beta^n u(f(k_n) - k_{n+1}) + \beta^{n+1} h(k_{n+1}).$$

Since $h \in S$, by taking the limits, we get $\lim_n T^n h(k_0) \geq V(k_0)$. If $k_0 = 0$, then $V(k_0) = -\infty$ and the previous inequality obviously holds.

We now prove that $\lim_n T^n h(k_0) \leq V(k_0)$. Fix some N. For any $n > N$, we find $k_1^n, k_2^n, ..., k_n^n$ with $g(k_t^n) \leq k_{t+1}^n \leq f(k_t^n)$ for $t = 0, 1, ..., n-1$, and

$$T^n h(k_0) = u(f(k_0) - k_1^n) + \beta u(f(k_1^n) - k_2^n) + ...$$
$$+ \beta^{N-1} u(f(k_{N-1}^n) - k_N^n) + \beta^N T^{n-N} h(k_N^n),$$
$$\leq u(f(k_0) - k_1^n) + \beta u(f(k_1^n) - k_2^n) + ...$$
$$+ \beta^{n-1} u(f(k_{N-1}^n) - k_N^n) + \beta^N h(k_N^n).$$

As before, the sequence $(k_0, k_1^n, ..., k_n^n, g(k_n^n), g(k_n^n), ...)$ may be assumed to converge to $\mathbf{k} \in \Pi(k_0)$ when n converges to infinity. Therefore:

$$\lim_n T^n h(k_0) \leq u(f(k_0) - k_1) + \beta u(f(k_1) - k_2) + ...$$

$$+\beta^{N-1} u(f(k_{N-1}) - k_N) + \beta^N h(k_N),$$

since $\limsup_n h(k_N^n) \leq h(k_N)$ (h is upper semi-continuous). Let N converge to infinity. Then

$$\lim_n T^n h(k_0) \leq U(\mathbf{k}) + \limsup_N \beta^N h(k_N) \leq U(\mathbf{k}) \leq V(k_0).$$

(iii) From (ii), it suffices to prove that such a function h belongs to S. It remains to prove that for any $k_0 > 0$, for any $\mathbf{k} \in \Pi'(k_0)$, we have $\lim_t \beta^t h(k_t) = 0$. But assumption (c) implies that for any k_0, we have $V(k_0) \leq h(k_0)$. Hence, for $\mathbf{k} \in \Pi'(k_0)$, we have $V(k_t) \leq h(k_t), \forall t$. Therefore

$$0 = \lim_t \beta^t V(k_t) \leq \liminf_t \beta^t h(k_t).$$

Since $\limsup_t \beta^t h(k_t) \leq 0$, we obtain the desired result. □

5.4 Optimal Policy - Properties of Optimal Paths

Proposition 5.4.1. *Assume that the assumptions of Proposition 5.3.3 hold. Let G be the optimal correspondence, i.e.*

$$\forall k \geq 0, G(k) = argmax\{u(f(k) - y) + \beta V(y) : g(x) \leq y \leq f(x)\}.$$

Then G is an upper semi-continuous correspondence.

Proof. We cannot use the Maximum Theorem because V equals $-\infty$ at the origin. We give a direct proof.
Let $\{k_0^n\}$ be a sequence that converges to k_0 and let $y^n \in G(k_0^n), \forall n$. We have $g(k_0^n) \leq y^n \leq f(k_0^n), \forall n$. Since f and g are continuous, the sequence $\{y^n\}$ will be uniformly bounded when n is large enough.

We can assume that it converges to some $y \in [g(k_0), f(k_0)]$. Let $z \in [g(k_0), f(k_0)]$. It is easy to check that one can find a sequence $\{z^n\}$ included in $[g(k_0^n), k_0^n]$ for every n large enough, which converges to z. We have:

$$\forall n, V(k_0^n) = u(f(k_0^n) - y^n) + \beta V(y^n)$$
$$\geq u(f(k_0^n) - z^n) + \beta V(z^n).$$

Let n go to infinity. Then

$$V(k_0) = u(f(k_0) - y) + \beta V(y) \geq u(f(k_0) - z) + \beta V(z).$$

Thus $y \in G(k_0)$. □

Proposition 5.4.2. *Assume* **H1, H2, H3, H4**. *Then:*
(i) A sequence $\mathbf{k}^* \in \Pi(k_0)$ *is optimal iff*
(a) $\forall t, k_{t+1}^* \in G(k_t^*)$,
(b) $\forall t, V(k_t^*) = u(f(k_t^*) - k_{t+1}^*) + \beta V(k_{t+1}^*)$.
(ii) Let $k_0 > 0$ *and let* $\mathbf{k}^*$ *be an optimal path. Then*
(a) $\forall t, 0 < k_{t+1}^* < f(k_t^*)$ *and*
(b) If $g(k_t^*) < k_{t+1}^*, \forall t \geq 0$, *then Euler equation holds:*

$$\forall t, u'(f(k_t^*) - k_{t+1}^*) = \beta u'(f(k_{t+1}^*) - k_{t+2}^*) f'(k_{t+1}^*).$$

(iii) Assume that f *is increasing and* g *is non-decreasing. In this case,* G *is non-decreasing (if* $g = 0$, *then* G *is increasing). Any optimal path is monotonic. If* $f'(0) > \frac{1}{\beta}$, *and* $k_0 \neq 0$, *then it is increasing.*
(iv) In addition to assumptions **H1, H2, H3, H4**, *assume that* f *is strictly concave and* g *is convex. Then the optimal solution is unique.*
If a sequence $\mathbf{k} \in \Pi(k_0)$ *satisfies the following conditions*
(a) $\forall t, g(k_t) < k_{t+1} < f(k_t)$,
(b) Euler equation, for every t,
(c) Transversality condition: $\lim_t \beta^t u'(f(k_t) - k_{t+1}) k_{t+1} = 0$,
then $\mathbf{k}$ *is the optimal solution.*

If moreover $f'(+\infty) < 1, g = 0$, *then* $\mathbf{k} \in \Pi(k_0)$ *is optimal iff it satisfies (b) and (c).*

Proof. (i) The proof is easy.
(ii) If $k_{t+1}^* = f(k_t^*) = 0$ for some t, then $V(k_0) = -\infty$ in contradiction with **H4**. If $k_t^* = 0$ for some t, then $k_{t+1}^* = 0$ and $V(k_0) = -\infty$: again, a contradiction. Hence, $\forall t, 0 < k_{t+1}^* < f(k_t^*)$.
The proof of (b) is the same as in Chapter 1.
(iii) Use the proof given in Chapter 3.
(iv) Use the proofs given in Chapter 1. □

5.5 Examples

5.5.1 Example 1

Consider the model

$$\max \sum_{t=0}^{+\infty} \beta^t Log(c_t), \ 0 < \beta < 1$$

$$\forall t, \ c_t + k_{t+1} \leq k_t^{\alpha}, \ 0 < \alpha < 1$$

$$c_t \geq 0, \ k_t \geq 0$$

$$k_0 \geq 0 \ \text{ is given.}$$

This model is equivalent to

$$\max \sum_{t=0}^{+\infty} \beta^t Log(k_t^{\alpha} - k_{t+1}), \ 0 < \beta \leq 1$$

$$\forall t, \ 0 \leq k_{t+1} \leq k_t^{\alpha}, \ 0 \leq \alpha < 1$$

$$k_0 \geq 0 \ \text{ is given.}$$

We check that this model satisfies **H1, H2, H3, H4**. We have $f(k) = k^{\alpha}$, $g(k) = 0$, $\Pi(k_0) = \{\mathbf{k} \mid 0 \leq k_{t+1} \leq k_t^{\alpha}\}$. Observe that 1 satisfies $1 = f(1)$. Moreover if $0 < k < 1$ then $k < f(k) < 1$ and if $k > 1$ then $f(k) < k$. Hence, $0 \leq y \leq f(k) \Rightarrow y \leq k + 1$.
H1 is satisfied (in particular $\gamma = 1, \gamma' = 1$), **H2** and **H3** are obviously satisfied. To check **H4**, one can easily prove that for any $k_0 > 0$ there exists $\hat{k}$ such that $\hat{k} < 1$, $\hat{k} < f(k_0)$ and $\hat{k} < f(\hat{k})$.

Hence for any $k_0 > 0$, the sequence $\mathbf{k} = (k_0, \hat{k}, \hat{k}, ...)$ is feasible Obviously,

$$\sum_{t=0}^{+\infty} \beta^t u(f(k_t) - k_{t+1}) = u(f(k_0) - \hat{k}) + \frac{\beta}{1-\beta} u(f(\hat{k}) - \hat{k}) > -\infty.$$

Since the utility function (which is logarithmic) is strictly concave, and the production function f is convex and increasing, from Proposition 5.4.2 (iv), there exists a unique optimal solution $\mathbf{k}$. Furthermore, since $g = 0$, f is increasing, from statement (ii) and (iii) of the same proposition, the optimal path $\mathbf{k}$ is increasing and satisfies Euler equation. In this model, it will converge to the steady state k^s which satisfies $f'(k^s) = \frac{1}{\beta}$ (the proof is the same as in chapter 1).

We now prove that the Value function V is continuous. We use Proposition 5.3.3. For that, we show that there exists a continuous function ϕ such that $V(k_0) \geq \phi(k_0), \forall k_0$ and if $\mathbf{k} \in \Pi'(k_0)$, then $\lim_t \beta^t \phi(k_t) = 0$.
Take some $\hat{k} \in]0, 1[$. Define

$$\phi(k_0) = \frac{Log(k_0^\alpha - k_0)}{1-\beta} \quad \text{if} \;\; 0 < k_0 < \hat{k}$$

$$\phi(k_0) = Log(k_0^\alpha - \hat{k}) + \frac{\beta}{1-\beta} Log(\hat{k}^\alpha - \hat{k}), \;\; \text{if} \;\; k_0 \geq \hat{k}.$$

Obviously ϕ is continuous. Assume $k_0 \geq \hat{k}$. If $k_0 \geq 1$, then $\hat{k} < 1 = f(1) \leq f(k_0)$. Since $\hat{k} < f(\hat{k})$, the sequence $\mathbf{k} = (k_0, \hat{k}, \hat{k}, ...)$ is feasible. If $\hat{k} \leq k_0 < 1$ then $\hat{k} \leq k_0 \leq f(k_0)$ and again the sequence $\mathbf{k} = (k_0, \hat{k}, \hat{k}, ...)$ is feasible. We have: $V(k_0) \geq \phi(k_0) = \sum \beta^t Log(f(k_t) - k_{t+1})$.
We now prove that, for any $\mathbf{k} \in \Pi'(k_0)$, $\lim_t \beta^t \phi(k_t) = 0$.
Since $k_t \leq k_0^{\alpha^t}$ for any t, we have

$$-\infty < \sum \beta^t Log(k_t^\alpha - k_{t+1}) \leq \sum \beta^t Log(k_t^\alpha)$$

$$\leq \sum (\alpha\beta)^t Log(k_0) < +\infty$$

and hence $\lim_t \beta^t Log(k_t) = 0$.

We have two cases: either there exists $0 < \tilde{k} < \hat{k}$ such that $k_t \geq \tilde{k} > 0$ for every t, or there exists a subsequence $k_\nu \longrightarrow 0$. In the first case, we have either

$$\min_{\tilde{k} \leq k \leq \hat{k}} \phi(k) \leq \phi(k_t) \leq \frac{\alpha Log(k_t)}{1-\beta}$$

if $k_t \leq \hat{k}$
or

$$\phi(\hat{k}) \leq \phi(k_t) \leq \alpha Log(k_t) + \frac{\beta}{1-\beta} Log(\hat{k}^\alpha - \hat{k}),$$

if $k_t > \hat{k}$,
and thus $\lim_t \beta^t \phi(k_t) = 0$.
In the second case, we have:

$$\beta^\nu \phi(k_\nu) = \beta^\nu \frac{Log(k_\nu^\alpha - k_\nu)}{1-\beta} = \frac{\beta^\nu}{1-\beta}(Log(k_\nu)^\alpha - Log(1 - k_\nu^{1-\alpha})).$$

Since $k_\nu \longrightarrow 0$, we have $\lim_\nu \beta^\nu \phi(k_\nu) = 0$.
We have proved that V is continuous.
We now want to compute V and the optimal policy. For that, we will apply Proposition 5.3.5(iii). Define $h(k) = \frac{\alpha}{1-\alpha\beta} Log(k)$. Since for any $\mathbf{k} \in \Pi(k_0)$ we have $k_t \leq k_0^{\alpha^t}, \forall t$, then $V(k_0) \leq h(k_0)$ for any $k_0 \geq 0$. We can find by direct calculation that

$$Th(k_0) = h(k_0) + Log(1-\alpha\beta) + \frac{\alpha\beta}{1-\alpha\beta} Log(\alpha\beta) < h(k_0).$$

Since for any $\mathbf{k} \in \Pi(k_0)$, we have $k_t \leq k_0^{\alpha^t}$, then:

$$\limsup_t \beta^t h(k_t) \leq \lim_t \frac{\beta^t \alpha^{t+1}}{1-\alpha\beta} Log(k_0) = 0.$$

Hence, $V = \lim_n T^n h$. Tedious computations yield

$$T^n h(k_0) = h(k_0) + \frac{1-\beta^n}{1-\beta}[Log(1-\alpha\beta) + \frac{\alpha\beta}{1-\alpha\beta} Log(\alpha\beta)]$$

and then

$$V(k_0) = h(k_0) + \frac{1}{1-\beta}[Log(1-\alpha\beta) + \frac{\alpha\beta}{1-\alpha\beta}Log(\alpha\beta)].$$

To compute the optimal policy we use the Bellman equation

$$V(k_0) = \max_{0 \le y \le f(k_0)} \{Log(k_0^\alpha - y) + \beta V(y)\}.$$

We can easily find the optimal value $k_1 = \alpha\beta k_0^\alpha$.

5.5.2 Example 2: The AK model

We consider the model

$$\max \sum \beta^t \frac{c_t^\theta}{\theta},\ \theta < 0$$

$$\begin{aligned} \forall t,\ c_t + I_t \le Ak_t,\ 0 < A \\ I_t = k_{t+1} - (1-\delta)k_t \\ c_t \ge 0,\ k_t \ge 0,\ I_t \ge 0 \\ k_0 \ge 0 \text{ is given.} \end{aligned}$$

We suppose $(1-\delta) < \beta(A+1-\delta) < 1$ and $A+1-\delta > 1$. This model is equivalent to

$$\max \sum \beta^t \frac{((A+1-\delta)k_t - k_{t+1})^\theta}{\theta},\ \theta < 0$$

$$\begin{aligned} (1-\delta)k_t \le k_{t+1} \le (A+1-\delta)k_t \\ k_0 \ge 0 \text{ is given.} \end{aligned}$$

The reader can check that the assumptions **H1, H2, H3, H4** are satisfied. To prove that the Value function V is continuous we observe that for any $k_0 \ge 0$ we have $V(k_0) \ge \phi(k_0) = \frac{(A-\delta)^\theta}{\theta}\frac{1}{1-\beta}k_0^\theta$. We want to show that for any $k > 0$ for any $\mathbf{k} \in \Pi'(k_0)$, $\lim_t \beta^t\phi(k_t) = 0$, or equivalently $\lim_t \beta^t k_t^\theta = 0$. But, for $\mathbf{k} \in \Pi'(k_0)$ we have

$$-\infty < \sum \beta^t \frac{((A+1-\delta)k_t - k_{t+1})^\theta}{\theta} \le \sum \beta^t \frac{A^\theta {k_t}^\theta}{\theta} < 0.$$

Hence $\lim_t \beta^t k_t^\theta = 0$. Apply Proposition 5.3.2 to conclude that V is continuous.
The reader can check that the sequence $\mathbf{k}^*$ defined by $k_t^* = \lambda^t k_0$ for every t and where $\lambda = (\frac{1}{\beta(A+1-\delta)})^{\frac{1}{\theta-1}}$ is optimal (Proposition 5.4.2(ii)).

6. Optimal Growth and Competitive Equilibrium

In this chapter we will prove that there exists an equivalence between the existence of an optimal path in the Ramsey model and the existence of a competitive equilibrium in an intertemporal model. For simplicity, we consider an one agent, one sector model.When investment is irreversible, i.e. constrained to be positive at every period, the reader can refer to Aliprantis, Border and Burkinshaw [2] or Duran and Le Van [36]. For existence of competitive equilibrium in an intertemporal setting with many agents and many sectors, one can refer, e.g., to Bewley [14] where utilities are expressed in terms of preorders, or Bewley [15] where utilities are additively separable, and Dana and Le Van [26] where utilities are represented by recursive preferences.

6.1 The Model

Consider again the Ramsey Model

$$\max \sum_{t=0}^{+\infty} \beta^t u(c_t)$$

under the constraints

$$\forall t \geq 0, c_t + k_{t+1} \leq f(k_t), k_t \geq 0, c_t \geq 0, \; k_0 > 0 \text{ is given}$$

As in the previous chapter, we introduce the following assumptions:
H1 $0 < \beta < 1$.
H2 The function $u: \mathbb{R}_+ \to \mathbb{R}_+$ is strictly concave, differentiable and satisfies $u' > 0, \; u(0) = 0$.
H3 The utility function u satisfies Inada condition $u'(0) = +\infty$.

H4 The production function $f: \mathbb{R}_+ \to \mathbb{R}_+$ is strictly concave, differentiable and satisfies $f' > 0,\ f(0) = 0, \lim_{x\to+\infty} f'(x) < 1$.

Under these assumptions, we know, from chapter 2, that there exists a unique optimal sequence $(\mathbf{c}^*, \mathbf{k}^*)$ verifying $\forall t, c_t^* > 0,\ k_t^* > 0$. Moreover, the sequence $\mathbf{k}^*$ is monotonic. If $f'(0) > \frac{1}{\beta}$, then the optimal sequence $\mathbf{k}^*$ converges to a steady state $k^s > 0$, and if $f'(0) \leq \frac{1}{\beta}$, then it converges to zero.
The purpose of this chapter is to prove that any competitive equilibrium of an intertemporal economy is the solution to the Ramsey model and, under an additional assumption, the solution to the Ramsey model is a competitive equilibrium.
Let us define our intertemporal economy and define a competitive equilibrium of this economy. In this economy, the sequences of consumptions and of capitals will be uniformly bounded, i.e. $\sup_{t=0,1,..,+\infty} c_t < +\infty,\ \sup_{t=0,1,..,+\infty} k_t < +\infty$. The sequence of prices $\mathbf{p}$ will satisfy $\sum_{t=0}^{+\infty} p_t < +\infty$. We say that the consumption and capital paths belong to

$$l_+^\infty = \{\mathbf{x} \in (\mathbb{R}_+)^\infty : \sup_t x_t < +\infty\},$$

and the sequences of prices belong to

$$l_+^1 = \{\mathbf{p} \in (\mathbb{R}_+)^\infty : \sum_{t=0}^{+\infty} p_t < +\infty\}.$$

In our economy, there is one consumer who lives an infinite number of periods. He owns the initial capital stock k_0 and owns the firm. There is one firm which produces at every date t, through a production function f, one good which is used as consumption good and capital good as well.

A competitive equilibrium of this economy is a list $\{\mathbf{c}^*, \mathbf{k}^*, \mathbf{p}^*, q^*\}$ such that
(i) $\mathbf{c}^* \in l_+^\infty, \mathbf{k}^* \in l_+^\infty, \mathbf{p}^* \in l_+^1 \backslash \{0\}, q^* \in \mathbb{R}_{++}$,
(ii) $\mathbf{c}^*$ is a solution to the consumer's problem:

$$\max \sum_{t=0}^{+\infty} \beta^t u(c_t)$$

under the constraint:

$$\sum_{t=0}^{+\infty} p_t^* c_t \leq q^* k_0 + \pi,$$

where π is the maximum profit of the firm,
(iii) $\mathbf{k}^*$ is a solution to the firm's problem

$$\pi = \max_{\mathbf{k}} \sum_{t=0}^{+\infty} p_t^* (f(k_t) - k_{t+1}) - q^* k_0,$$

under the constraints:

$$\forall t \geq 0,\ 0 \leq k_{t+1} \leq f(k_t),\ k_0 > 0 \text{ is given},$$

(iv) and the market clears at every period:

$$\forall t \geq 0,\ c_t^* + k_{t+1}^* = f(k_t^*),\ k_0^* = k_0.$$

The proof will be done in two stages. At a first stage, we assume Inada condition. This assumption will be dropped at the second stage.

We introduce an additional assumption:
H5 The production function satisfies $f'(0) > 1$.

We have the first result:

Proposition 6.1.1. *(i) Assume* **H1-H4**. *We have:*
If $\{\mathbf{c}^*, \mathbf{k}^*, \mathbf{p}^*, q^*\}$ *is a competitive equilibrium, then* $\{\mathbf{c}^*, \mathbf{k}^*\}$*solve the Ramsey problem.*
(ii) Assume **H1-H5**. *Let* $\{\mathbf{c}^*, \mathbf{k}^*\}$ *solve the Ramsey problem. Define* $q^* = 1$, *and* $\forall t \geq 0,\ p_t^* = \beta^t u'(c_t^*)$. *Then* $\{\mathbf{c}^*, \mathbf{k}^*, \mathbf{p}^*, q^*\}$ *is a competitive equilibrium.*

Proof. (i) Let $\{\mathbf{c}^*, \mathbf{k}^*, \mathbf{p}^*, q^*\}$ be a competitive equilibrium and let $\{\mathbf{c}, \mathbf{k}\}$ be sequences of consumptions and of capitals which satisfy the constraints of the Ramsey model. From chapter 1, we know that there exists A such that $\forall t \geq 0,\ k_t \leq A,\ c_t \leq f(A)$. Hence $\sum_{t=0}^{+\infty} p_t^* c_t$, $\sum_{t=0}^{+\infty} p_t^* (f(k_t) - k_{t+1})$ are finite valued. Moreover,

$$\sum_{t=0}^{+\infty} p_t^* c_t \leq \sum_{t=0}^{+\infty} p_t^*(f(k_t) - k_{t+1}).$$

Since $\mathbf{k}^*$ solves the producer's problem, we have:

$$\sum_{t=0}^{+\infty} p_t^*(f(k_t) - k_{t+1}) \leq \sum_{t=0}^{+\infty} p_t^*(f(k_t^*) - k_{t+1}^*) = \pi + q^* k_0,$$

where π is the maximum profit. Hence

$$\sum_{t=0}^{+\infty} p_t^* c_t \leq \pi + q^* k_0$$

which implies

$$\sum_{t=0}^{+\infty} \beta^t u(c_t^*) \geq \sum_{t=0}^{+\infty} \beta^t u(c_t).$$

We have proved that $\{\mathbf{c}^*, \mathbf{k}^*\}$ solve the Ramsey problem.
(ii) We proceed in two steps. Let $\{\mathbf{c}^*, \mathbf{k}^*\}$ solve the Ramsey problem.
Step 1. Assume that $f'(0) > \frac{1}{\beta}$.
In this case, $k_t^* \to k^s > 0$ (see chapter 1), and $c_t^* \to c^s = f(k^s) - k^s > 0$. The sequence $\mathbf{p}^*$ with $p_t^* = \beta^t u'(c_t^*), \forall t$, belongs therefore to $l_+^1 \backslash \{0\}$. Let $q^* = 1$ and $\pi = \sum_{t=0}^{+\infty} p_t^*(f(k_t^*) - k_{t+1}^*) - q^* k_0$.
Let $\mathbf{c}$ satisfy

$$\sum_{t=0}^{+\infty} p_t^* c_t \leq q^* k_0 + \pi = \sum_{t=0}^{+\infty} p_t^*(f(k_t^*) - k_{t+1}^*).$$

Since

$$\sum_{t=0}^{+\infty} p_t^* c_t^* = \sum_{t=0}^{+\infty} p_t^*(f(k_t^*) - k_{t+1}^*) \geq \sum_{t=0}^{+\infty} p_t^* c_t,$$

by the concavity of u, we have:

$$\sum_{t=0}^{+\infty} \beta^t u(c_t^*) - \sum_{t=0}^{+\infty} \beta^t u(c^t) \geq \sum_{t=0}^{+\infty} \beta^t u'(c_t^*)(c_t^* - c_t)$$

$$= \sum_{t=0}^{+\infty} p_t^* c_t^* - \sum_{t=0}^{+\infty} p_t^* c_t \geq 0.$$

In other words, the sequence $\mathbf{c}^*$ solves the consumers's problem. We now prove that the sequence $\mathbf{k}^*$ maximizes the profit of the firm, given the prices $\mathbf{p}^*$.
Let

$$\Delta_T = \sum_{t=0}^{T} p_t^*(f(k_t^*) - k_{t+1}^*) - \sum_{t=0}^{T} p_t^*(f(k_t) - k_{t+1}),$$

where the sequence $\mathbf{k}$ satisfies $\forall t,\ 0 \leq k_{t+1} \leq f(k_t)$. We have, by the concavity of f:

$$\Delta_T \geq \sum_{t=0}^{T} p_t^* f'(k_t^*)(k_t^* - k_t) - \sum_{t=0}^{T} p_t^*(k_{t+1}^* - k_{t+1})$$

$$= [p_1^* f'(k_1^*) - p_0^*](k_1^* - k_1) + ...$$

$$+[p_T^* f'(k_T^*) - p_{T-1}^*](k_T^* - k_T) - p_T^*(k_{T+1}^* - k_{T+1}).$$

For $t = 1, ..., T$, we have from the Euler equations:

$$p_t^* f'(k_t^*) - p_{t-1}^* = \beta^t u'(c_t^*) f'(k_t^*) - \beta^{t-1} u'(c_{t-1}^*) = 0.$$

Hence

$$\Delta_T \geq -p_T^* k_{T+1}^* + p_T^* k_{T+1} \geq -p_T^* k_{T+1}^*.$$

Since $k_{T+1}^* \to k^s$ and $p_T^* = \beta^T u'(c_T^*) \to 0$ when $T \to +\infty$, we have $\lim_{T\to+\infty} \Delta_T \geq 0$. We have proved that $\mathbf{k}^*$ solves the firm's problem.
Finally, the market clears at every period, since $\forall t,\ c_t^* + k_{t+1}^* = f(k_t^*)$.
Summing up, $\{\mathbf{c}^*, \mathbf{k}^*, \mathbf{p}^*, 1\}$ is a competitive equilibrium.
Step 2. We now assume that $f'(0) \leq \frac{1}{\beta}$.
In this case we know that the sequence $\mathbf{k}^*$ converges to zero. From **H5**, there exists $\hat{k}$ satisfying $f'(\hat{k}) > 1$. There exists T such that $k_t^* \leq \hat{k},\ \forall t \geq T$. The Euler equations hold for $t \geq T$:

$$\beta^t u'(c_t^*) = \beta^{t+1} u'(c_{t+1}^*) f'(k_{t+1}^*).$$

Since $k_{t+1}^* \leq \hat{k}$, we have

$$\beta^t u'(c_t^*) \geq \beta^{t+1} u'(c_{t+1}^*) f'(\hat{k}),$$

or

$$\beta^{t+1} u'(c_{t+1}^*) \leq \frac{1}{f'(\hat{k})} \beta^t u'(c_t^*).$$

Let $\gamma = \frac{1}{f'(\hat{k})}$. Then

$$\forall \tau \geq 1, \; \beta^{t+\tau} u'(c_{t+\tau}^*) \leq \gamma^\tau \beta^T u'(c_T^*).$$

Since $\gamma < 1$, we get:

$$\sum_{t=0}^{+\infty} \beta^t u'(c_t^*) = \sum_{t=0}^{T} \beta^t u'(c_t^*) + \frac{\gamma}{1-\gamma} \beta^T u'(c_T^*) < +\infty.$$

That proves that $\mathbf{p}^* \in l_+^1 \setminus \{0\}$.
To prove that $\{\mathbf{c}^*, \mathbf{k}^*, \mathbf{p}^*, 1\}$ is a competitive equilibrium, proceed as in Step 1. □

We will now drop assumption **H3** (Inada assumption). To prove the existence of an equilibrium, we will take a sequence of utility functions which satisfy Inada condition and which converges pointwise to the initial utility function. With each utility function of the sequence one can associate, from Proposition 6.1.1, an equilibrium. We then prove that this sequence of equilibria converges to an equilibrium of the initial economy.

Proposition 6.1.2. *Assume* **H1, H3, H4, H5** *and that* $u'(0) < +\infty$. *Then:*
(i) If $\{\mathbf{c}^*, \mathbf{k}^*, \mathbf{p}^*, q^*\}$ *is a competitive equilibrium, then* $\{\mathbf{c}^*, \mathbf{k}^*\}$ *is the solution to the Ramsey model.*
(ii) Conversely, if $\{\mathbf{c}^*, \mathbf{k}^*\}$ *is the solution to the Ramsey model, then there exists* $\mathbf{p}^* \in l_+^1 \setminus \{0\}$ *such that* $\{\mathbf{c}^*, \mathbf{k}^*, \mathbf{p}^*, 1\}$ *is a competitive equilibrium.*

Proof. (i) The proof is the same as in Proposition 6.1.1.
(ii) Let $u_n(c) = u(c) + \frac{1}{n}\frac{c^\theta}{\theta}$, where $\theta \in]0, 1[$. Consider the economy with u_n as utility function and f as technology.
For simplicity, we only consider the case with $f'(0) > \frac{1}{\beta}$. The Ramsey model with u_n as instantaneous utility function admits

an optimal path $\mathbf{k_n^*}$ which converges to the steady state k^s (which does not depend on n, since $f'(k^s) = \frac{1}{\beta}$). From Proposition 6.1.1, $\{\mathbf{c_n^*}, \mathbf{k_n^*}, \mathbf{p_n^*}, 1\}$ with $p_{n,t}^* = \beta^t(u'(c_{n,t}^*) + \frac{1}{n}(c_{n,t}^*)^{\theta-1}), \forall t$, is a competitive equilibrium.

We will show that the sequence $\{u_n'(c_{n,t}^*)\}$ is uniformly bounded from above when n converges to ∞.

First suppose that $0 < k_0 < k^s$. Let $\bar{k}$ satisfy $\bar{k} = f(\bar{k})$. The sequence $\{f^t(k_0)\}_{t=1,\dots,+\infty}$ converges to $\bar{k}$. Since $\bar{k} > k^s$, there exists T such that $f^{T+1}(k_0) > k^s$. Fix such a T. For any n, the sequence $\mathbf{k_n}$ is increasing. Since $k_0 < k^s$, we have $k_0 < k_{n,t}^* < k^s, \forall t \geq 1,\ \forall n$, since the optimal policy is increasing, k^s is optimal from itself and $k_{n,t}^*$ is optimal from $k_{n,t-1}^*$.

From the Euler equation

$$u_n'(c_{n,t}^*) = \beta u_n'(c_{n,t+1}^*) f'(k_{n,t+1}^*),$$

one obtains $u_n'(c_{n,t}^*) > u_n'(c_{n,t+1}^*),\ \forall t,\ \forall n$.

Hence, $u_n'(c_{n,t}^*) < u_n'(c_{n,0}^*),\ \forall t,\ \forall n$. It suffices to prove that the sequence $\{u'(c_{n,0}^*)\}_n$ is uniformly bounded from above. On the contrary, there will be a subsequence that, without loss of generality, we will denote by $\{u'(c_{n,0}^*)\}_n$, which converges to infinity. From the Euler equation:

$$u'(c_{n,0}^*) = \beta u'(c_{n,1}^*) f'(k_{n,1}^*)$$

we obtain that $u^*(c_{n,1}')$ converges to infinity since $f'(k^s) < f'(k_{n,1}^*) < f'(k_0)$.

By induction, $u^*(c_{n,t}')$ converges to infinity, with $t = 2, \dots, T$. That implies that $c_{n,0}^*, c_{n,1}^*, \dots, c_{n,T}^*$ converges to zero. Since we have $\forall t, c_{n,t}^* + k_{n,t+1}^* = f(k_n^*)$, then $k_{n,1}^*$ will converge to $f(k_0)$, $k_{n,2}^*$ will converge to $f^2(k_0)$, etc..., and $k_{n,T+1}^*$ will converge to $f^{T+1}(k_0)$. Since $f^{T+1}(k_0) > k^s$, we have a contradiction with $k_{n,t}^* < k^s, \forall t,\ \forall n$.

If $k_0 = k^s$, then $u_n'(c_{n,t}^*) = u_n'(c^s), \forall n,\ \forall t$, and hence, converges to $u'(c^s)$ when n converges to infinity. That means that the sequence $\{u_n'(c_{n,t}^*)\}$ is uniformly bounded from above.

Now consider the case where $k_0 > k^s$. The sequence $\mathbf{k_n^*}$ is decreasing and converges to k^s. We have: $\forall t,\ k_0 > k_{n,t}^* > k^s$ and $f'(k_{n,t}^*) < \frac{1}{\beta}$. Hence, $c_{n,t}^* > c_{n,t+1}^*, \forall t,\ \forall n$. The sequence $\mathbf{c_n^*}$ is decreasing and converges to c^s. Therefore, $\forall n,\ \forall t, u_n'(c_{n,t}^*) < u_n'(c^s)$. The sequence

$\{u'_n(c^*_{n,t})\}$ is uniformly bounded from above since $u'_n(c^s)$ converges to $u'(c^s)$.

To sum up, we have proved that there exists M such that $u'_n(c^*_{n,t}) \leq M$, $\forall n$, $\forall t$. Since there exists $A > 0$ such that $c^*_{n,t} \leq A, \forall n$, $\forall t$, we have

$$\forall n,\ \forall t, \beta^t u'(A) \leq \beta^t u'_n(c^*_{n,t}) \leq \beta^t M.$$

Therefore, one can assume that the sequence of prices $\mathbf{p^*_n} = \{(\beta^t u'_n(c^*_{n,t})\}_n$ converges pointwise to $\mathbf{p_0} \in l^1_+ \backslash \{0\}$.

Let $\mathbf{c_0}, \mathbf{k_0}$ be the limits pointwise of $\mathbf{c^*_n}, \mathbf{k^*_n}$. We will prove that $(\mathbf{c_0}, \mathbf{k_0}, \mathbf{p_0}, 1)$ is a competitive equilibrium.

First, the market clears at every date, since

$$\forall n, \forall t,\ c^*_{n,t} + k^*_{n,t+1} = f(k^*_{n,t})$$

and therefore

$$\forall t,\ c_{0,t} + k_{0,t+1} = f(k_{0,t}).$$

Now consider the consumer's problem. Let $\mathbf{c}$ satisfy:

$$\sum_{t=0}^{+\infty} p_{0,t} c_t \leq \sum_{t=0}^{+\infty} p_{0,t}[f(k_{0,t}) - k_{0,t+1}].$$

Observe that

$$\forall n,\ \forall T,\ \sum_{t=0}^{T} p^*_{n,t} c^*_{n,t} = \sum_{t=0}^{T} p^*_{n,t}[f(k^*_{n,t}) - k^*_{n,t+1}].$$

By taking the limits when n goes to infinity, we get:

$$\sum_{t=0}^{T} p_{0,t} c_{0,t} = \sum_{t=0}^{T} p_{0,t}[f(k_{0,t}) - k_{0,t+1}].$$

Now, take the limits when T converges to ∞:

$$\sum_{t=0}^{+\infty} p_{0,t} c_{0,t} = \sum_{t=0}^{+\infty} p_{0,t}[f(k_{0,t}) - k_{0,t+1}] < +\infty.$$

Consider the inequality:

$$\sum_{t=0}^{T} \beta^t u_n(c^*_{n,t}) - \sum_{t=0}^{T} \beta^t u_n(c_t) \geq \sum_{t=0}^{T} \beta^t u'(c^*_{n,t})(c^*_{n,t} - c_t).$$

Let n and T go to infinity. We obtain:

$$\sum_{t=0}^{+\infty} \beta^t u(c_{0,t}) - \sum_{t=0}^{+\infty} \beta^t u(c_t) \geq \sum_{t=0}^{+\infty} p_{0,t}(c_{0,t} - c_t)$$

$$= \sum_{t=0}^{+\infty} p_{0,t}[f(k_{0,t}) - k_{0,t+1}] - \sum_{t=0}^{+\infty} p_{0,t}c_t \geq 0.$$

We have proved that the sequence $\mathbf{c_0}$ maximizes the intertemporal utility of the consumer under the budget constraint.
Let us prove that the sequence $\mathbf{k_0}$ maximizes the profit of the firm, given the prices $\mathbf{p_0}$.
Let $\mathbf{k}$ satisfy the feasibility constraints

$$\forall t,\ 0 \leq k_{t+1} \leq f(k_t).$$

Let

$$\Delta_{n,T} = \sum_{t=0}^{T} p^*_{n,t}(f(k^*_{n,t}) - k^*_{n,t+1}) - \sum_{t=0}^{T} p^*_{n,t}(f(k_t) - k_{t+1}).$$

Using the Euler equation, we obtain, as in the proof of Proposition 6.1.1, that $\Delta_{n,T} \geq -p^*_{n,T}k^*_{n,T+1}$.
Let n converge to ∞. Then

$$\Delta_{0,T} \geq -p_{0,T}k_{0,T+1}.$$

Since $0 \leq p_{0,T} \leq M,\ k_{0,T+1} \leq A, \forall T$, we have $\lim_T \Delta_{0,T} \geq 0$, i.e.:

$$\sum_{t=0}^{+\infty} p_{0,t}(f(k_{0,t}) - k_{0,t+1}) - \sum_{t=0}^{+\infty} p_{0,t}(f(k_t) - k_{t+1}) \geq 0.$$

We have proved that $\mathbf{k_0}$ solves the producer's problem.
To end the proof, observe that from statement (i), $(\mathbf{c_0}, \mathbf{k_0})$ is an optimal solution to the Ramsey problem. Since this problem has a unique solution, any optimal path must coincide with $(\mathbf{c_0}, \mathbf{k_0})$. □

Remark 6.1.1. (i) Assumption **H5** is equivalent to the Adequacy Assumption in Bewley [14], [15] or the Interiority Assumption in Dana and Le Van [26]. It implies that the production set has an interior point. Hence, there exists a sustainable production sequence. We will point out that, in our model, this assumption is crucial to have equilibrium prices in l^1_+.
Indeed, assume that Inada condition holds. Let $(\mathbf{c}^*, \mathbf{k}^*, \mathbf{p}^*, q^*)$ be an equilibrium. We know, from the consumer's programme, that the equilibrium prices will be $p^*_t = \lambda\beta^t u'(c^*_t), \forall t$, where λ is a positive constant. If $f'(0) \leq 1$, then $f'(k^*_t) < 1, \forall t$, since f' is decreasing and $k^*_t > 0$. From the Euler equation, we have:

$$\beta^t u'(c^*_t) < \beta^{t+1} u'(c^*_{t+1}), \ \forall t \geq 0.$$

Hence $p^*_t > p^*_0$, $\forall t \geq 1$. The sequence of prices cannot belong to l^1_+.
(ii) We would like to mention that solving the Ramsey model is equivalent to solving a Pareto-optimal problem for the intertemporal economy which just has one consumer. We know, from the First Welfare Theorem, that any equilibrium is a Pareto-optimum. For our economy, any competitive equilibrium is a solution to the Ramsey model.
We also know, from the Second Welfare Theorem, that, for an economy with convex technology, under some interiority assumption, any Pareto-optimum is an equilibrium with transfer. When the economy has one agent, an equilibrium with transfer is an equilibrium of the economy. For our economy, a solution to the Ramsey model, under assumption **H5**, is an equilibrium.

7. Optimal Growth Models Without Discounting

There have been, as far as we are aware, two main approaches to the infinite time horizon optimal growth problem. The first is to discount future consumptions utilities by some arbitrary factor and then to maximize the resulting infinite series. This approach is used in the previous chapters. The second approach pioneered by Ramsey [66] and reworked by Samuelson and Solow [73], Koopmans [44] and von Weizsacker [83] uses a discount factor equal to one. Gale [39], and Brock [19] developed two criteria based on the overtaking criterion and characterized the optimality of an infinite sequence of capital stocks. of von Weizsacker [83]. Their proofs of existence of optimal paths used the fact that these paths converge to a stationary point. Gale and Brock did not use a dynamic programming approach. Dana and Le Van [25], [27] introduced the Value function of Gales's problem and showed that the optimal policy is obtained from a Bellman equation. The present chapter is based on the works of Gale [39] and Dana and Le Van [25]. It is organized as follows. The model is presented in Section 1. In Section 2, we define a good feasible sequence and give its properties. In particular, any good feasible sequence converges to a stationary point. This property is crucial to prove existence of optimal solutions, and to study the properties of the Value function and the optimal policy. The proof of the existence of optimal solutions and the properties of these solutions are presented in Section 3. In Section 4, we study the Value function and the Bellman equation. In particular, we prove that the Value function, as expected, is the unique solution, in some class of functions, to the Bellman equation.

7.1 The Model

We consider an intertemporal economy where the instantaneous utility of the representative consumer depends on k_t, the capital stock on hand at date t and on k_{t+1}, the capital stock for date $t+1$. Given k_t, the set of feasible capital stocks for the next period $t+1$ is $\Gamma(k_t)$. We assume that at any period t, the feasible capital stock on hand belongs to X, a subset of $\mathbb{R}^n_+$. More explicitly, we make the following assumptions:
H1: X is a compact, convex set of $\mathbb{R}^n_+$ with non-empty interior and X contains 0.
H2: Γ is a continuous correspondence from X into X with non-empty convex images. Its graph, denoted by $graph\Gamma$, is convex. Let us recall that $graph\Gamma = \{(x, y) \in X \times X : y \in \Gamma(x)\}$.
H3: (Free disposal assumption) If $y \in \Gamma(x), x' \geq x$ and $y' \leq y$, then $y' \in \Gamma(x')$.
H4: (Existence of expansible capital stocks) There exist $(x, y) \in graph\Gamma$, with $y >> x$, i.e. $y_i > x_i, \forall i = 1, ..., n$.
H5: The instantaneous utility function $F : graph\Gamma \to \mathbb{R}$ is strictly concave, continuous. It is increasing with respect to the first variable and decreasing with respect to the second variable.

Remark 7.1.1. Assumption **H3** implies that $0 \in \Gamma(0)$. Assumptions **H3** and **H4** imply that the interior of $graph\Gamma$, denoted by $int(graph\Gamma)$ is non-empty.

A sequence $\mathbf{x}$ is feasible from $x_0 \in X$ if, $\forall t \geq 0$, $x_{t+1} \in \Gamma(x_t)$. We denote by $\Pi(x_0)$ the set of feasible sequences from x_0. Following Gale (1967), a *programme* from x_0 is a feasible sequence from x_0. The set of programmes is denoted by Π, i.e. $\Pi = \cup_{x \in X} \Pi(x)$.

7.2 Good programmes

Let us introduce the optimal stationary programme.
Define

$$\bar{F} = \max\{F(x, y) : (x, y) \in graph\Gamma, y \geq x\}.$$

Assumption **H4** implies that the set $\{(x, y) \in graph\Gamma, y \geq x\}$ is non-empty. Since $graph\Gamma$ is compact, this set is compact. Therefore $\bar{F}$ exists in $\mathbb{R}$, since F is continuous. But, observe that from **H5**, this problem is equivalent to

$$\max\{F(x, x) : (x, x) \in graph\Gamma\}.$$

Proposition 7.2.1. *Assume* **H1-H5**. *Then:*
(i) there exists $\bar{x}$ *which satisfies:*

$$\bar{F} = F(\bar{x}, \bar{x}) = \max\{F(x, x) : (x, x) \in graph\Gamma\}.$$

(ii) there exists $\bar{p} \in \mathbb{R}^n_+$ *such that:*

$$F(x, y) + \bar{p} \cdot y - \bar{p} \cdot x \leq F(\bar{x}, \bar{x}), \ \forall (x, y) \in graph\Gamma.$$

Proof. (i) The existence of $\bar{x}$ is trivial.
(ii) The statement is just the Kuhn-Tucker condition to the problem:

$$\max\{F(x, y) : (x, y) \in graph\Gamma, y \geq x\},$$

Indeed, define $F_0(x, y) = F(x, y)$ if $(x, y) \in graph\Gamma$, and $F_0(x, y) = -\infty$ if $(x, y) \notin graph\Gamma$. The problem becomes:

$$\max\{F_0(x, y) : y \geq x\}.$$

Then apply Corollary 7.2.1 in Florenzano, Le Van and Gourdel [37].
□

We want that the stationary point $(\bar{x}, \bar{x})$ is unique. We now add an assumption which ensures the uniqueness of the stationary point.
H6: The utility function F is strictly concave at $(\bar{x}, \bar{x})$, i.e.

$$\forall (x, y) \neq (\bar{x}, \bar{x}), \forall \lambda \in]0, 1[,$$

$$F(\lambda x + (1 - \lambda)\bar{x}, \lambda y + (1 - \lambda)\bar{x}) > \lambda F(x, y) + (1 - \lambda) F(\bar{x}, \bar{x}).$$

Proposition 7.2.2. *Assume* **H1-H6**. *Then the stationary point* $\bar{x}$ *is unique.*

Proof. Assume there exists another solution $\bar{x}'$ to the problem $\max\{F(x,x) : (x,x) \in graph\Gamma\}$. Let $\lambda \in]0,1[$.
Since $graph\Gamma$ is convex $(\lambda\bar{x} + (1-\lambda)\bar{x}', \lambda\bar{x} + (1-\lambda)\bar{x}')$ belongs to $graph\Gamma$.
Since F is strictly concave at $(\bar{x},\bar{x})$, we have:

$$F(\lambda\bar{x}+(1-\lambda)\bar{x}', \lambda\bar{x}+(1-\lambda)\bar{x}') > \lambda F(\bar{x},\bar{x}) + (1-\lambda)F(\bar{x}',\bar{x}') = \bar{F}.$$

We have a contradiction. □

Theorem 7.2.1. *Assume* **H1-H6**. *For any programme* **x**, *either*
(i) $\lim_{T\to+\infty}\sum_{t=0}^{T}[F(x_t,x_{t+1}) - F(\bar{x},\bar{x})]$ *exists in* $\mathbb{R}$ *and* x_t *converges to* $\bar{x}$.
or (ii) $\lim_{T\to+\infty}\sum_{t=0}^{T}[F(x_t,x_{t+1}) - F(\bar{x},\bar{x})] = -\infty$.

Proof. Let $\delta(x_t,x_{t+1}) = F(\bar{x},\bar{x}) - F(x_t,x_{t+1}) + \bar{p}\cdot x_t - \bar{p}\cdot x_{t+1}$. By Proposition 7.2.1, $\delta(x_t,x_{t+1}) \geq 0$. Hence $\sum_{t=0}^{+\infty}\delta(x_t,x_{t+1})$ exists in $\mathbb{R}\cup\{+\infty\}$.
If $\sum_{t=0}^{+\infty}\delta(x_t,x_{t+1})$ is finite then $\delta(x_t,x_{t+1}) \to 0$ when $t \to +\infty$.
Let $\phi(x,y) = F(x,y) - \bar{p}\cdot x + \bar{p}\cdot y$. This function ϕ is strictly concave and have a unique maximizer which is $(\bar{x},\bar{x})$. Since $\{(x_t,x_{t+1})\}$ belongs to $graph\Gamma$ which is compact, we take a cluster point (x,x') of this sequence. Since $\delta(x_t,x_{t+1}) \to 0$, we have $\phi(x_t,x_{t+1}) \to F(\bar{x},\bar{x}) = \phi(\bar{x},\bar{x})$. Since ϕ is continuous, we have $\phi(x,x') = \phi(\bar{x},\bar{x})$. The maximizer being unique we have $(x,x') = (\bar{x},\bar{x})$. The sequence $\{(x_t,x_{t+1})\}$ has a unique cluster point. It is therefore convergent (see Appendix) and converges to $(\bar{x},\bar{x})$, or equivalently, $x_t \to \bar{x}$.
Since

$$\lim_{T\to+\infty}\sum_{t=0}^{T}[F(x_t,x_{t+1}) - F(\bar{x},\bar{x})] = -\sum_{t=0}^{+\infty}\delta(x_t,x_{t+1}) + p\cdot(x_0 - \bar{x}),$$

$\lim_{T\to+\infty}\sum_{t=0}^{T}[F(x_t,x_{t+1}) - F(\bar{x},\bar{x})]$ exists in $\mathbb{R}$.
If $\sum_{t=0}^{+\infty}\delta(x_t,x_{t+1})$ is infinite, as

$$\sum_{t=0}^{T}[F(x_t,x_{t+1}) - F(\bar{x},\bar{x})] = -\sum_{t=0}^{T}\delta(x_t,x_{t+1}) + p\cdot(x_0 - x_{T+1}),$$

and since **x** is bounded, we have

$$\limsup_{T \to +\infty} \sum_{t=0}^{T} [F(x_t, x_{t+1}) - F(\bar{x}, \bar{x})] \leq$$

$$-\sum_{t=0}^{+\infty} \delta(x_t, x_{t+1}) + \limsup_{T} [p \cdot (x_0 - x_{T+1})] = -\infty.$$

□

From the previous theorem, We may now introduce a definition of a good programme.
A programme $\mathbf{x} \in \Pi$ is *good* if $\lim_{T \to +\infty} \sum_{t=0}^{T} [F(x_t, x_{t+1}) - F(\bar{x}, \bar{x})]$ exists in $\mathbb{R}$.
A programme $\mathbf{x} \in \Pi$ is not good iff $\lim_{T \to +\infty} \sum_{t=0}^{T} [F(x_t, x_{t+1}) - F(\bar{x}, \bar{x})] = -\infty$.
An equivalent definition of good programmes is given by the following proposition.

Proposition 7.2.3. *Assume* **H1-H6**. *A programme* $\mathbf{x} \in \Pi$ *is good iff there exists* N *such that*

$$\forall T, \ \sum_{t=0}^{T} [F(x_t, x_{t+1}) - F(\bar{x}, \bar{x})] \geq N.$$

Proof. (i) Let $\mathbf{x}$ be a good programme. Let

$$\alpha = \lim_{T \to +\infty} \sum_{t=0}^{T} [F(x_t, x_{t+1}) - F(\bar{x}, \bar{x})].$$

Let $\alpha' < \alpha$. There exists T_0 such that for all $T \geq T_0$,

$$\sum_{t=0}^{T} [F(x_t, x_{t+1}) - F(\bar{x}, \bar{x})] \geq \alpha'.$$

Define $S_\tau = \sum_{t=0}^{\tau} [F(x_t, x_{t+1}) - F(\bar{x}, \bar{x})]$.
Take $N = \min\{S_0, S_1, ..., S_{T_0}, \alpha'\}$ to obtain that

$$\forall T, \ \sum_{t=0}^{T} [F(x_t, x_{t+1}) - F(\bar{x}, \bar{x})] \geq N.$$

(ii) Conversely, it follows from Theorem 7.2.1 that, if there exists N such that

$$\forall T, \sum_{t=0}^{T}[F(x_t, x_{t+1}) - F(\bar{x}, \bar{x})] \geq N,$$

then we cannot be in case (ii) of Theorem 7.2.1. □

The reader can check that our definition of good programmes is equivalent to the Gale's definition [39]:

A programme $\mathbf{x}$ is good if for any programme $\mathbf{x}'$ we have

$$\liminf_{T\to+\infty} \sum_{t=0}^{T}[u(x_t, x_{t+1}) - u(x'_t, x'_{t+1})] > -\infty.$$

We now give sufficient conditions for non emptiness of the set of good programmes from x_0.

Proposition 7.2.4. *Assume* **H1-H6** *and that there exists a neighborhood $\mathcal{U}$ of $(\bar{x}, \bar{x})$ in graphΓ, and $\sigma > 0$ such that*

$$|F(x, y) - F(\bar{x}, \bar{x})| \leq \sigma[\|x - \bar{x}\| + \|y - \bar{x}\|].$$

Then $\Pi_G(x_0)$ is non-empty for any x_0 such that $\Gamma(x_0)$ contains a strictly positive y_0 (i.e., all the components in $\mathbb{R}^n$ of y_0 are strictly positive.)

Proof. By **H4**, there exists (x_p, y_p) in *graph*Γ such that $x_p << y_p$. Since $0 \in \Gamma(0)$, we have $\lambda y_p \in \Gamma(\lambda x_p)$, for $\lambda \in]0, 1[$. If λ is sufficiently small, we have $\lambda y_p << y_0$ and $\lambda y_p >> \lambda x_p$. So, we can assume $y_p << y_0$ and $y_p >> x_p$.
Let $\lambda \in]0, 1[$. For any integer n, we remark that

$$(1 - \lambda^{n+1})\bar{x} + \lambda^{n+1} x_p \leq (1 - \lambda^n)\bar{x} + \lambda^n y_p \text{ iff } (1 - \lambda)\bar{x} + \lambda x_p \leq y_p.$$

Choose λ sufficiently close to 1 so that $(1 - \lambda)\bar{x} + \lambda x_p \leq y_p << y_0$. Define, for $t \geq 1$, $x_t = (1 - \lambda^t)\bar{x} + \lambda^t x_p$. By the choice of λ and by **H3**, we have $x_1 \in \Gamma(x_0)$. For $t \geq 1$,

$$x_{t+1} \leq (1 - \lambda^t)\bar{x} + \lambda^t y_p \in \Gamma((1 - \lambda^t)\bar{x} + \lambda^t x_p) = \Gamma(x_t).$$

By **H3**, $x_{t+1} \in \Gamma(x_t)$.
Since $x_t \to \bar{x}$, there exists T_0 such that for any $t \geq T_0$, $(x_t, x_{t+1}) \in \mathcal{U}$ and hence

$$|F(x_t, x_{t+1}) - F(\bar{x}, \bar{x})| \leq 2\sigma\lambda^t \|\bar{x} - x_p\|.$$

Thus,

$$\sum_{t=T_0}^{+\infty} |F(x_t, x_{t+1}) - F(\bar{x}, \bar{x})| \leq 2\sigma \frac{\lambda^{T_0}}{1-\lambda} \|\bar{x} - x_p\|,$$

and $\mathbf{x}$ is good from x_0. □

Let $\Pi_G(x_0)$ denote the set of good programmes starting from $x_0 \in X$.

7.3 Optimal Programmes

We will now define an optimal programme for the optimal growth models without discounting.
A programme $\mathbf{x}^* \in \Pi(x_0)$ is optimal if, for any programme $\mathbf{x} \in \Pi(x_0)$, we have

$$\lim_{T \to +\infty} \sum_{t=0}^{T} [F(x_t^*, x_{t+1}^*) - F(\bar{x}, \bar{x})] \geq \lim_{T \to +\infty} \sum_{t=0}^{T} [F(x_t, x_{t+1}) - F(\bar{x}, \bar{x})]$$

From Theorem 7.2.1, this definition makes sense. Observe also that, if $\Pi_G(x_0)$ is non-empty, then an optimal programme is good.

Remark 7.3.1. Gale [39] gives another definition: A programme $\mathbf{x}^* \in \Pi(x_0)$ is optimal if, for any programme $\mathbf{x} \in \Pi(x_0)$, we have

$$\limsup_{T \to +\infty} \sum_{t=0}^{T} [F(x_t, x_{t+1}) - F(x_t^*, x_{t+1}^*)] \leq 0.$$

One can check that an optimum in the sense of Gale is optimal with our definition. When $\Pi_G(x_0)$ is non-empty, the two definitions coincide. Another definition is given by Brock [19]:

A programme $\mathbf{x}^* \in \Pi(x_0)$ is *weakly* optimal if, for any $\mathbf{x} \in \Pi(x_0)$, we have:

$$\liminf_{T\to+\infty} \sum_{t=0}^{T} [F(x_t, x_{t+1}) - F(x_t^*, x_{t+1}^*)] \leq 0$$

The reader who is interested by optimal programmes in the sense of Brock, can refer to Brock [19] or Dana and Le Van [27].

We first show that the stationary programme $(\bar{x}, \bar{x}, ..., \bar{x}, ...)$ is an optimal programme.

Proposition 7.3.1. *Assume* **H1-H6**. *Then the stationary programme* $(\bar{x}, \bar{x}, ..., \bar{x}, ...)$ *is an optimal programme from* $\bar{x}$.

Proof. Let $\mathbf{x} \in \Pi(\bar{x})$. One can easily check that

$$\sum_{t=0}^{T} [F(x_t, x_{t+1}) - F(\bar{x}, \bar{x})] \leq \bar{p} \cdot \bar{x} - \bar{p} \cdot x_{t+1}.$$

If $\mathbf{x}$ is good, then $x_t \to \bar{x}$ (Theorem 7.2.1). Hence,

$$\lim_{T} \sum_{t=0}^{T} [F(x_t, x_{t+1}) - F(\bar{x}, \bar{x})] \leq 0 = \sum_{t=0}^{+\infty} [F(\bar{x}, \bar{x}) - F(\bar{x}, \bar{x})].$$

If $\mathbf{x}$ is not good, then $\sum_{t=0}^{+\infty} [F(x_t, x_{t+1}) - F(\bar{x}, \bar{x})] = -\infty$ (Theorem 7.2.1).
The proof is complete. □

From Theorem 7.2.1, the function

$$\gamma : \Pi \to [-\infty, +\infty[, \text{ defined by}$$

$$\gamma(\mathbf{x}) = \lim_{T\to+\infty} \sum_{t=0}^{T} [F(x_t, x_{t+1}) - F(\bar{x}, \bar{x})]$$

is well-defined.

Proposition 7.3.2. *Assume* **H1-H6**. *Then the function* γ *is concave, upper semi-continuous for the product topology.*

Proof. For $\mathbf{x} \in \Pi$, define

$$\delta(x_t, x_{t+1}) = F(\bar{x}, \bar{x}) - F(x_t, x_{t+1}) + \bar{p} \cdot x_t - \bar{p} \cdot x_{t+1}.$$

The function δ is non-negative, concave and continuous for the product topology. We have

$$\sum_{t=0}^{T} [F(x_t, x_{t+1}) - F(\bar{x}, \bar{x})] = \sum_{t=0}^{T} -\delta(x_t, x_{t+1}) + \bar{p} \cdot x_0 - \bar{p} \cdot x_{T+1}.$$

We claim that:

$$\gamma(\mathbf{x}) = \lim_{T \to +\infty} \Psi_T(\mathbf{x}) + \bar{p} \cdot x_0 - \bar{p} \cdot \bar{x},$$

where $\Psi_T(\mathbf{x}) = \sum_{t=0}^{T} -\delta(x_t, x_{t+1})$.
When $\mathbf{x}$ is good, then $x_T \to \bar{x}$ and the claim is true. When $\mathbf{x}$ is not good, then $\sum_{t=0}^{+\infty} [F(x_t, x_{t+1}) - F(\bar{x}, \bar{x})] = -\infty$. Since the sequence $\mathbf{x}$ is bounded, $\Psi_T(\mathbf{x})$ converges to $-\infty$ and the claim is also true.
For every T, Ψ_T is concave, continuous for the product topology and $\Psi_T \geq \Psi_{T+1}$. Since $\gamma(\mathbf{x})$ is the decreasing limit of continuous functions, it is upper semi-continuous (see Appendix). It is concave as limit of concave functions. □

Our optimal growth model turns out to be:

$$\max\{\gamma(\mathbf{x}) : \mathbf{x} \in \Pi(x_0)\}.$$

We obtain the following existence of solution result.

Theorem 7.3.1. *Assume* **H1-H6**.
Let $x_0 \in X$. Then there exists an optimal solution from x_0. If furthermore $\Pi_G(x_0) \neq \emptyset$, then any optimal solution from x_0 is good.

Proof. For the product topology, the function γ is upper semi-continuous and the set $\Pi(x_0)$ is compact, non-empty. There thus exists an optimal solution (see Appendix). The second statement is obvious. □

If we assume that the utility function u is continuously differentiable, then any interior good programme is optimal iff it satisfies the Euler equation.

Proposition 7.3.3. *Assume* **H1-H6** *and that u is continuously differentiable. Assume also that $\Pi_G(x_0) \neq \emptyset$.*
Let $\mathbf{x}$ *be a programme from x_0 which satisfies $\forall t$, $(x_t, x_{t+1}) \in int(graph\Gamma)$. Then* $\mathbf{x}$ *is optimal iff it satisfies the Euler equation:*

$$\forall t,\ F_2(x_t, x_{t+1}) + F_1(x_{t+1}, x_{t+2}) = 0.$$

Proof. Since $\Pi_G(x_0) \neq \emptyset$, any optimal programme must be good. The proof that, if $\mathbf{x}$ is good and optimal, then it satisfies the Euler equation, is the same as in chapter 4, Proposition 4.1.5.
Let us prove the converse. Assume that $\mathbf{x}$ satisfies the Euler equation. We first prove that it is a good programme. Let $\mathbf{x}'$ be another good programme from x_0. Let

$$\Delta_T = \sum_{t=0}^{T}[F(x_t, x_{t+1}) - F(x'_t, x'_{t+1})].$$

We have:

$$\Delta_T \geq \sum_{t=0}^{T}[F_1(x_t, x_{t+1}).(x_t - x'_t) + F_2(x_t, x_{t+1}).(x_{t+1} - x'_{t+1})$$

$$= (F_2(x_0, x_1) + F_1(x_1, x_2)).(x_1 - x'_1) + ...+$$

$$(F_2(x_{T-1}, x_T)+F_1(x_T, x_{T+1})).(x_T-x'_T)+F_2(x_T, x_{T+1}).(x_{T+1}-x'_{T+1})$$

$$\geq F_2(x_T, x_{T+1}).(x_{T+1} - x'_{T+1}) \geq F_2(x_T, x_{T+1}).x_{T+1} \geq N,$$

since we have

$$\|F_2(x_T, x_{T+1})\| \leq \max\{\|F_2(x, y)\| : (x, y) \in graph\Gamma\}$$

and since x_{T+1} belongs to X which is compact.
Observe that

$$\Delta_T = \sum_{t=0}^{T}[F(x_t, x_{t+1}) - F(\bar{x}, \bar{x})] - \sum_{t=0}^{T}[F(x'_t, x'_{t+1}) - F(\bar{x}, \bar{x})].$$

Since $\mathbf{x}'$ is a good programme, from Proposition 7.2.3, there exists N' such that $\sum_{t=0}^{T}[F(x'_t, x'_{t+1}) - F(\bar{x}, \bar{x})] \geq N'$. Hence,

$$\sum_{t=0}^{T}[F(x_t, x_{t+1}) - F(\bar{x}, \bar{x})] \geq N + N'.$$

From Proposition 7.2.3, the sequence $\mathbf{x}$ is good. Since $\mathbf{x}$ and $\mathbf{x}'$ are good, $x_{T+1} \to \bar{x}$, $x'_{T+1} \to \bar{x}$.
Hence $\lim_T \Delta_T \geq 0$ and

$$\sum_{t=0}^{+\infty}[F(x_t, x_{t+1}) - F(\bar{x}, \bar{x})] \geq \sum_{t=0}^{+\infty}[F(x'_t, x'_{t+1}) - F(\bar{x}, \bar{x})].$$

That proves that $\mathbf{x}$ is optimal. □

7.4 Value Function - Bellman Equation

We now introduce as in the previous chapters the Value function of our optimal growth model. Let V denote the Value function, i.e.:

$$\forall x_0 \in X, V(x_0) = \sup\{\gamma(\mathbf{x}) : \mathbf{x} \in \Pi(x_0)\}.$$

Observe that if $\Pi_G(x_0) \neq \emptyset$, then $V(x_0) > -\infty$ and

$$V(x_0) = \max\{\gamma(\mathbf{x}) : \mathbf{x} \in \Pi_G(x_0)\}.$$

If $\Pi_G(x_0) = \emptyset$, then $V(x_0) = -\infty$.

Proposition 7.4.1. *Assume* **H1-H6**. *Then: (i) The Value function V is concave, non-decreasing, upper semi-continuous and satisfies the Bellman equation:*

$$(\mathcal{B}) \qquad \forall x \in X, V(x) = \sup\{F(x,y) - F(\bar{x}, \bar{x}) + V(y) : y \in \Gamma(x)\}.$$

(ii) Moreover, $V(\bar{x}) = 0$ *and,*

$$\forall \mathbf{x} \in \Pi_G(x_0), \lim_{T \to +\infty} V(x_T) = 0.$$

Proof. (i)It is easy to check that V is concave.
To check that V is non-decreasing, observe that, from assumption **H3**, any sequence $\mathbf{x} \in \Pi(x_0)$ belongs also to $\Pi(x'_0)$ if $x_0 \leq x'_0$. Hence, $\gamma(\mathbf{x}) \leq V(x'_0), \forall \mathbf{x} \in \Pi(x_0)$. That implies $V(x_0) \leq V(x'_0)$.

To prove that V is upper semi-continuous, take a sequence $\{x_0^n\}$ converging to x_0. Let $a = \limsup_n V(x_0^n)$, and let $\{x_0^{n_k}\}$ satisfy $\lim_k V(x_0^{n_k}) = a$.

If $a = -\infty$, we have $\limsup_n V(x_0^n) \leq V(x_0)$. If $a > -\infty$, we can assume $V(x_0^{n_k}) > -\infty$, $\forall k$. Therefore, there exists $\mathbf{x^{n_k}} \in \Pi(x_0^{n_k})$ such that $V(x_0^{n_k}) = \gamma(\mathbf{x^{n_k}})$. Since $\mathbf{x^{n_k}}$ belongs to X^∞ which is compact for the product topology, one may assume that $\mathbf{x^{n_k}}$ converges to some $\mathbf{x}$. It is easy to see that $\mathbf{x} \in \Pi(x_0)$. Since γ is upper semi-continuous, we have:

$$a = \lim_k \gamma(\mathbf{x^{n_k}}) \leq \gamma(\mathbf{x}) \leq V(x_0).$$

We have proved the upper semi-continuity of V.

We now prove that V satisfies the Bellman equation.

If $V(x) > -\infty$, then $\Pi_G(x)$ is non-empty. From Theorem 7.3.1, there exists an optimal programme starting from x. By the usual argument,

$$V(x) = \max\{F(x,y) - F(\bar{x},\bar{x}) + V(y) : y \in \Gamma(x)\}.$$

If $V(x) = -\infty$, then $V(y) = -\infty$, for every $y \in \Gamma(x)$ (if not, there would exist a good programme from x), and the Bellman equation holds.

(ii) It is obvious that $V(\bar{x}) = 0$.

Let $\mathbf{x} \in \Pi_G(x_0)$. We know that $x_t \to \bar{x}$ when $t \to +\infty$. Since V is upper semi-continuous, $\limsup_T V(x_T) \leq V(\bar{x}) = 0$.

Let us show that $\liminf_T V(x_T) \geq 0$.

For any T we have:

$$-\infty < \gamma(\mathbf{x}) = \sum_0^{+\infty}[F(x_t, x_{t+1}) - F(\bar{x},\bar{x})]$$

$$= \sum_0^{T}[F(x_t, x_{t+1}) - F(\bar{x},\bar{x})] + \sum_{T+1}^{+\infty}[F(x_t, x_{t+1}) - F(\bar{x},\bar{x})]$$

$$\leq \sum_0^{T}[F(x_t, x_{t+1}) - F(\bar{x},\bar{x})] + V(x_{T+1}),$$

or

$$\gamma(\mathbf{x}) - \sum_0^T [F(x_t, x_{t+1}) - F(\bar{x}, \bar{x})] \leq V(x_{T+1}).$$

Taking the limits, we obtain: $0 \leq \liminf_T V(x_{T+1})$. □

We want to prove that the Value function is the unique solution to the Bellman equation in some class of functions.

Let S be the set of upper semi-continuous functions W from X into $\mathbb{R} \cup \{-\infty\}$ which satisfy $\lim_t W(x_t) = 0$ for any $\mathbf{x} \in \Pi_G(x_0)$, for any $x_0 \in X$.

Theorem 7.4.1. *Assume* **H1-H6**. *Then V is a unique solution to the Bellman equation in S.*

Proof. Let W be another solution to the Bellman equation in S. Let $x_0 \in X$.
Assume $W(x_0) > -\infty$. We first prove that $W(x_0) \leq V(x_0)$. Since W is upper semi-continuous, u is continuous, $\Pi(x_0)$ is compact, there exists $x_1 \in \Pi(x_0)$ such that

$$W(x_0) = F(x_0, x_1) - F(\bar{x}, \bar{x}) + W(x_1).$$

By induction, we find $(x_2, x_3, ..., x_T, ...)$ such that $\forall t, x_{t+1} \in \Pi(x_t)$, and $W(x_t) = F(x_t, x_{t+1}) - F(\bar{x}, \bar{x}) + W(x_{t+1})$. That implies:

$$\forall T,\ W(x_0) = \sum_0^T [F(x_t, x_{t+1}) - F(\bar{x}, \bar{x})] + W(x_{T+1}).$$

Since W is upper semi-continuous and X is compact, $\sup\{W(x) : x \in X\} = M < +\infty$.
Hence $W(x_0) - M \leq \sum_0^T [F(x_t, x_{t+1}) - F(\bar{x}, \bar{x})], \forall T$. From Proposition 7.2.3, the sequence $(x_0, x_1, ..., x_T, ...)$ is good. Thus:

$$W(x_0) = \lim_T \sum_0^T [F(x_t, x_{t+1}) - F(\bar{x}, \bar{x})] + \lim_T W(x_{T+1})$$

$$= \gamma(\mathbf{x}) \leq V(x_0).$$

Let us prove that $W(x_0) \geq V(x_0)$.

For any $\mathbf{x} \in \Pi_G(x_0)$, we have

$$W(x_0) \geq \sum_0^T [F(x_t, x_{t+1}) - F(\bar{x}, \bar{x})] + W(x_{T+1}),$$

and hence

$$W(x_0) \geq \lim_T \sum_0^T [F(x_t, x_{t+1}) - F(\bar{x}, \bar{x})] + \lim_T W(x_{T+1}) = \gamma(\mathbf{x}).$$

Since $V(x_0) = \max\{\gamma(\mathbf{x}) : (\mathbf{x}) \in \Pi_G(x_0)\}$, we have $W(x_0) \geq V(x_0)$. Observe that if $\Pi_G(x_0) \neq \emptyset$, then $W(x_0) > -\infty$. Indeed, take some $\mathbf{x} \in \Pi_G(x_0)$. We have

$$\forall T, W(x_0) \geq \sum_0^T [F(x_t, x_{t+1}) - F(\bar{x}, \bar{x})] + W(x_{T+1}),$$

and therefore,

$$W(x_0) \geq \lim_T \sum_0^T [F(x_t, x_{t+1}) - F(\bar{x}, \bar{x})] + \lim_T W(x_{T+1})$$

$$= \sum_0^{+\infty} [F(x_t, x_{t+1}) - F(\bar{x}, \bar{x})] > -\infty.$$

When $W(x_0) = -\infty$, we have $\Pi_G(x_0) = \emptyset$. In this case, we have $V(x_0) = -\infty$.
We have proved that $W = V$. □

Let us add another assumption.
H7: The stationary point $(\bar{x}, \bar{x}) \in int(graph\Gamma)$.

Corollary 7.4.1. *Assume* **H1-H6** *and* **H7**. *Then V is the unique upper semi-continuous solution to the Bellman equation which satisfies $V(\bar{x}) = 0$.*

Proof. Let W be another solution which is upper semi-continuous and satisfies $W(\bar{x}) = 0$. Let $\mathbf{x} \in \Pi_G(x_0)$. Then $x_t \to \bar{x}$. We have $\limsup_t W(x_t) \leq W(\bar{x}) = 0$.

Since $(x_t, x_{t+1}) \rightarrow (\bar{x}, \bar{x})$, from **H7**, $\bar{x} \in \Gamma(x_t)$ when t is large enough. We have:

$$W(x_t) \geq F(x_t, \bar{x}) - F(\bar{x}, \bar{x}) + W(\bar{x}) = F(x_t, \bar{x}) - F(\bar{x}, \bar{x}),$$

and

$$\liminf_t W(x_t) \geq \lim_t F(x_t, \bar{x}) - F(\bar{x}, \bar{x}) = 0.$$

Thus, $\lim_t W(x_t) = 0$.
We have proved that W belongs to S. Apply Theorem 7.4.1. □

Let $dom(V)$ denote the domain of V, i.e.

$$dom(V) = \{x \in X : V(x) > -\infty\}.$$

Since V is concave, it is continuous in the interior of $dom(V)$ (see the Appendix). We want to show that V is continuous on $dom(V)$ under two sets of assumptions (see Theorem 7.4.2, Proposition 7.4.2). Moreover, it is continuous in a generalized sense: if $x \notin dom(V)$,and if $x^n \rightarrow x$, then $V(x^n) \rightarrow V(x) = -\infty$.

Theorem 7.4.2. *Assume* **H1-H7**. *Then V is a unique solution to the Bellman equation which is upper semi-continuous and satisfies $V(\bar{x}) = 0$.*
Moreover, V is continuous in a generalized sense.

Proof. From Corollary 7.4.1, we know that V is the unique solution to the Bellman equation which is upper semi-continuous and satisfies $V(\bar{x}) = 0$.
Let us prove the last claim. We distinguish two cases: (i) V is continuous on $dom(V)$, (ii) $V(x^n) \rightarrow -\infty$ when $x^n \rightarrow x \notin dom(V)$.
Case (ii) is easy.
Indeed, since V is upper semi-continuous, $\limsup_n V(x^n) \leq V(x) = -\infty$, when $x^n \rightarrow x \notin dom(V)$. This implies that $\liminf_n V(x^n) = \limsup_n V(x^n) = \lim_n V(x^n) = -\infty$.
Consider case (i).
Let $x_0 \in dom(V)$ and $\mathbf{x}$ be an optimal sequence from x_0. The sequence $\mathbf{x}$ is good and hence $x_t \rightarrow \bar{x}$. From assumption **H7**, there exists T_0 such that $\forall T \geq T_0$, $(x_T, \bar{x}) \in int(graph\Gamma)$. We therefore have:

$$V(x_0) \geq \sum_{t=0}^{T}[F(x_t, x_{t+1}) - F(\bar{x}, \bar{x})] + V(\bar{x}).$$

Fix some $T \geq T_0$.
Let $\{x_0^n\} \subset dom(V)$ converge to x_0. Since V is upper semi-continuous, to prove that V is continuous at x_0, it suffices to show that $\liminf_n V(x_0^n) \geq V(x_0)$.
Let $\{x_0^{n_k}\}$ be a subsequence of $\{x_0^n\}$ which satisfies $\lim_k V(x_0^{n_k}) = \liminf_n V(x_0^n)$. Since the correspondence Γ is lower semi-continuous, there exist sequences $\{x_1^{n_k}\} \subset X, ..., \{x_T^{n_k}\} \subset X$, such that:

$$\forall\ t = 1, ..., T+1,\ \forall k,\ x_t^{n_k} \in \Gamma(x_{t-1}^{n_k}),$$

and when $k \to +\infty$, $x_t^{n_k} \to x_t$, $\forall t = 0, ..., T+1$.
Since $(x_{T+1}, \bar{x}) \in int(graph\Gamma)$, for every k large enough, we have $(x_{T+1}^{n_k}, \bar{x}) \in graph\Gamma$. Therefore, for k large enough:

$$V(x_0^{n_k}) \geq \sum_{t=0}^{T}[F(x_t^{n_k}, x_{t+1}^{n_k}) - F(\bar{x}, \bar{x})] + (F(x_{T+1}^{n_k}, \bar{x}) - F(\bar{x}, \bar{x})) + V(\bar{x}).$$

Hence

$$\lim_k V(x_0^{n_k}) \geq \sum_{t=0}^{T}[F(x_t, x_{t+1}) - F(\bar{x}, \bar{x})] + F(x_{T+1}, \bar{x}) - F(\bar{x}, \bar{x}).$$

Let $T \to +\infty$. Then, we get $\liminf_n V(x_0^n) \geq V(x_0)$ since $x_{T+1} \to \bar{x}$. The proof is complete. □

We will now show that the optimal policy is continuous.

Theorem 7.4.3. *Assume* **H1-H7**. *For $x_0 \in dom(V)$, define the optimal policy g:*

$$g(x_0) = argmax\{F(x_0, y) + V(y) : V(y) \in \Gamma(x_0)\}.$$

Then g is a continuous mapping on $dom(V)$. The optimal programme from x_0, $\mathbf{x}$, satisfies $\forall t \geq 0, x_t = g^t(x_0)$, where g^t denotes the t-iterated of g.

Proof. Since F is strictly concave and V is concave, the *argmax* is single-valued. From Theorem 7.4.2, V is continuous and finite-valued on $dom(V)$. From the maximum theorem, g is continuous on $dom(V)$.
It is obvious that if $x_0 \in dom(V)$, then the optimal programme $\mathbf{x}$ satisfies $x_t = g^t(x_0),\ \forall t \geq 0$. □

We will show that the Value function may be differentiable.

Proposition 7.4.2. *Assume* **H1-H6**. *Let* $\mathbf{x}$ *be the optimal path from* $x_0 \in dom(V)$. *If* $(x_0, x_1) \in int(graph\Gamma)$, *then* V *is continuous at* x_0.
Assume moreover that F *is differentiable on* $int(graph\Gamma)$.
If $(x_0, x_1) \in int(graph\Gamma)$, *then* V *is differentiable at* x_0 *and* $V'(x_0) = F_1(x_0, x_1)$.

Proof. We have $V(x_0) = F(x_0, x_1) + V(x_1)$. Since $x_0 \in dom(V)$, we have $V(x_0) > -\infty$ and hence, $V(x_1) > -\infty$. If $(x_0, x_1) \in int(graph\Gamma)$, then for any x_0' in a neighborhood of x_0, we have $x_1 \in \Gamma(x_0')$. Hence:

$$V(x_0') \geq F(x_0', x_1) - F(\bar{x}, \bar{x}) + V(x_1) > -\infty.$$

That means that $x_0 \in int(dom(V))$. Since V is concave, it is continuous at x_0 (see Appendix).
The proof of the differentiability of V is exactly the same as in chapter 4. □

We now give three examples. The first is the one-sector Ramsey model with decreasing returns production function. In the second example, we have a Leontieff technology. The third example is the two-sector model.

Example 1 Let $X = [0,1]$. Let $f : [0,1] \rightarrow [0,1]$ be a continuously differentiable, strictly concave and increasing function. We assume $f'(0) > 1$. Define for $x \in X$, $\Gamma(x) = [0, f(x)]$, and for $(x, y) \in graph\Gamma$, $F(x, y) = v(f(x) - y)$, where v is a strictly concave function, increasing, $v(0) = 0$. The optimal stationary point $\bar{x}$ is unique $(f'(\bar{x}) = 1)$. Observe that $\bar{x}$ corresponds to the golden

rule, i.e., $f(\bar{x}) - \bar{x}$ is the maximal stationary consumption. Observe also that $(\bar{x}, \bar{x})$ is in the interior of $graph\Gamma$.
Assumptions **H1-H7** are satisfied. For any $x_0 > 0$, there exists a unique optimal programme. The Value function V is continuous in a generalized sense, and the optimal policy g is continuous in $]0, 1]$.

Example 2 Let us consider a Von Neumann economy. Let K be a convex, compact, non-empty set of $\mathbb{R}^k_+$ constraining the activity levels v. Let A be an $(n \times k)-$ goods input matrix and B be an $(n \times k)-$ goods output matrix. The matrices A, B are non-negative. Let X be a convex, compact set of $\mathbb{R}^n_+$ containing $A(K) \cup B(K)$ where

$$A(K) = \{z \in \mathbb{R}^n : z = Ax,\ x \in \mathbb{R}^k\},$$

and

$$B(K) = \{z \in \mathbb{R}^n : z = Bx,\ x \in \mathbb{R}^k\}.$$

The technology correspondence is:

$$\text{for } x \in X,\ \Gamma(x) = \{y \in X : \exists v \in K, x \geq Av \text{ and } y \leq Bv\}.$$

We assume that, for every i, there exists j such that $b_{ij} > 0$, i.e., all goods are producible.
For $(x, y) \in graph\Gamma$, define

$$F(x, y) = \max\{\tilde{u}(v) : Av \leq x, Bv \geq y\},$$

where $\tilde{u}$ is a strictly concave function from $\mathbb{R}^k_+$ into $\mathbb{R}$. Assume also there exists $v \in K$ such that $Bv >> Av$. We let to reader check that this model satisfies the assumptions **H1-H6**. Moreover, if $x >> 0$, then there exists $v >> 0$ such that $Av \leq x$ and $Bv >> 0$. Assume that the stationary point $(\bar{x}, \bar{x})$ is in the interior of $graph\Gamma$. In this case the utility function F is subdifferentiable at $(\bar{x}, \bar{x})$ (see Appendix). One can apply Proposition 7.2.4. That means that $\Pi_G(x) \neq \emptyset, \forall x >> 0$. Therefore, there exists an optimal solution for any $x >> 0$. The Value function V is continuous and the optimal policy is also continuous on $dom(V)$.

Example 3 We let to the reader check that the two-sectors model (Example 4) of chapter 4 satisfies the assumptions **H1-H7** of the present chapter.

8. Turnpike in Optimal Growth Models with Convex Technology

In chapter 2, we have shown that, in the Ramsey model, the optimal capital stocks converge to a strictly positive stock if the initial capital stock is strictly positive and if the marginal productivity at the origin of the production function is high enough.
In chapter 7, in a n-sector growth model with convex technology and where the future utility is not discounted, the optimal capital stocks converge to a positive steady if the initial capital stock is expansible.
In this chapter, we will show that this property still holds when the discount factor is close to one. The proof is based on the papers by Scheinkman [76] and by McKenzie [56]. It is divided in three steps. First, we prove that, given a neighborhood of the optimal steady state, there exists a value β_0, such that, if the discount factor β is in the interval $[\beta_0, 1[$, the optimal capital path will "visit" this neighborhood at some date T. Second, we prove that there exists a value β_1 such that, for any $\beta \geq \beta_1$, the optimal path will not escape from this neighborhood after some date T_1. Finally, we prove the asymptotic stability of the optimal path in this neighborhood.
Our strategy is as follows. We first prove the turnpike property for a compact technology set D (Theorem 8.3.1). This result will then allow us to obtain the turnpike theorem for unbounded technology sets (Theorem 8.3.2).
For simplicity, we will consider a one-sector model. The reader can check that our method can be extended to multi-sector models (for more details, see Scheinkman [76] and McKenzie [56]).

Let us consider the model:

$$\max \sum_{t=0}^{+\infty} \beta^t F(x_t, x_{t+1}),\ \beta \in [0, 1[,$$

under the constraints:

$$(x_t, x_{t+1}) \in D, \ \forall t \geq 0, \ x_0 \in \mathbb{R}_+ \text{ is given.}$$

We first introduce the following assumptions
H1 D is a convex, compact, non-empty set of $\mathbb{R}^2_+$.
More precisely

$$(x, y) \in D \Leftrightarrow (x, y) \in \mathbb{R}^2_+, y \leq f(x), x \leq a,$$

with $a > 0$, and f, a differentiable, strictly concave, increasing function. The function f satisfies: $f(0) = 0$, there exists $\hat{x}$ with $0 < \hat{x} < a$, $f(\hat{x}) = \hat{x}$.
H2 The function $F : D \to \mathbb{R}_+$ is twice continuously differentiable, strictly concave. It is increasing with respect to the first variable, decreasing with respect to the second variable.

8.1 The Visit Lemmas

These lemmas will be proved by a "cascade" of lemmas.

Lemma 8.1.1. *Under assumption* **H1**, *we have:*
(i) There exist $\gamma' > 0, 0 < \gamma < 1$, such that:

$$(x, y) \in D \Rightarrow y \leq \gamma x + \gamma',$$

(ii) There exist $R > 0, 0 < \hat{\gamma} < 1$, such that

$$x > R, (x, y) \in D \Rightarrow y \leq \hat{\gamma} x,$$

(iii) If $x > \hat{x}$, then $f(x) < x$.

Proof. (i) Since f is strictly concave and $f(0) = 0$, we have

$$f'(\hat{x}) < \frac{f(\hat{x})}{\hat{x}} = 1.$$

For any $x \geq 0$, we have by the concavity of f:

$$f(\hat{x}) - f(x) \geq f'(\hat{x})(\hat{x} - x),$$

and thus

$$f(x) \leq f(\hat{x}) - f'(\hat{x})\hat{x} + f'(\hat{x})x.$$

Take $\gamma = f'(\hat{x}) < 1$ and $\gamma' = f(\hat{x}) - f'(\hat{x})\hat{x} = (1 - f'(\hat{x}))\hat{x} > 0$.
(ii) Let $\hat{\gamma}$ satisfy $\gamma < \hat{\gamma} < 1$. Define $R = \frac{\gamma'}{(\hat{\gamma}-\gamma)}$. If $x > R$, then $(\hat{\gamma} - \gamma)x > \gamma'$. Hence, by (i), $(x, y) \in D$ implies $y \leq \hat{\gamma}x$.
(iii) If $x > \hat{x}$, then there exists $0 < \lambda < 1$ such that $\hat{x} = \lambda x$. By strict concavity, $f(\lambda x) = f(\lambda x + (1 - \lambda)0) > \lambda f(x)$. But $f(\lambda x) = f(\hat{x}) = \hat{x} = \lambda x$. Hence $f(x) < x$. □

Lemma 8.1.2 (Free Disposal). *Assume* **H1, H2**. *Let* $(x, y) \in D$. *If* $a \geq x' \geq x$, $y' \leq y$, *then* $(x', y') \in D$.

Proof. It is obvious. □

We recall that x is *expansible* if there exists y which satisfies $(x, y) \in D$ and $y > x$.

Lemma 8.1.3 (Existence of expansible capital stocks). *The set of expansible capital stocks is* $]0, \hat{x}[$.

Proof. It is obvious that if $x = 0$ or $x = \hat{x}$, then it is not expansible. From Lemma 8.1.1(iii), if $x > \hat{x}$, then it is not expansible.
If $x \in]0, \hat{x}[$, then there exists $\lambda \in]0, 1[$ such that $x = \lambda\hat{x}$. By strict concavity, $f(x) = f(\lambda\hat{x}) > \lambda f(\hat{x}) = \lambda\hat{x} = x$. Take $y = f(x)$ and conclude. □

Lemma 8.1.4. *Let* $x_0 \in]0, \hat{x}[, x \in]0, \hat{x}[, x_0 \neq x$. *Then there exists a finite sequence* $(x_1, x_2, ..., x_T)$ *such that* $(x_0, x_1) \in D, (x_1, x_2) \in D, ..., (x_T, x) \in D$.

Proof. First case: $x < x_0$. In this case we have $x < f(x) < f(x_0)$ and $(x_0, x) \in D$. The claim is true.
Second case: $x > x_0$. If $x \leq f(x_0)$, then $(x_0, x) \in D$ and we end the proof. Consider the case where $x > f(x_0)$. Observe that the sequence $\{f^n(x_0)\}$ is increasing and converges to $\hat{x}$. Hence, there exists T such that $f^T(x_0) \leq x \leq f^{T+1}(x_0)$. Define $x_t = f^t(x_0)$, $\forall t = 1, ..., T$. Then $(x_t, x_{t+1}) \in D, \forall t = 0, ..., T - 1$ and $(x_T, x) \in D$. □

We say that x is a stationary capital stock if $(x, x) \in D$. Let P denote the set of stationary capital stocks. Obviously, P is non-empty since 0 and $\hat{x}$ belong to P. More precisely, we have:

Lemma 8.1.5. *The set P is equal to the segment $[0, \hat{x}]$.*

Proof. It follows from Lemma 8.1.1(iii) that there exists no stationary stock above $\hat{x}$. Hence, $P \subset [0, \hat{x}]$.
Let $x \in]0, \hat{x}[$. It follows from Lemmas 8.1.3 and 8.1.2 that $(x, x) \in D$. That means that $]0, \hat{x}[\subset P$. Since 0 and $\hat{x}$ belong to P, the proof is established. □

Lemma 8.1.6. *There exists a unique $\bar{x}$ such that*

$$F(\bar{x}, \bar{x}) = \max\{F(x, x) : x \in P\}.$$

Proof. Since P is compact, there exists a solution $\bar{x}$. It is unique since F is strictly concave. □

We now add the following assumption:
H3 $(\bar{x}, \bar{x}) \in intD$.

Lemma 8.1.7. *There exists $\bar{p} > 0$ such that*

$$F(x, y) + \bar{p}y - \bar{p}x < F(\bar{x}, \bar{x}), \ \forall (x, y) \in D, \ \textit{and } (x, y) \neq (\bar{x}, \bar{x}).$$

Proof. Since $(\bar{x}, \bar{x})$ is in the interior of D, we have $F_1(\bar{x}, \bar{x}) + F_2(\bar{x}, \bar{x}) = 0$ (F_1, F_2 are the partial derivatives of F). By the strict concavity of F, we have:

$$F(\bar{x}, \bar{x}) - F(x, y) > \bar{p}(\bar{x} - x) - \bar{p}(\bar{x} - y) = \bar{p}(y - x),$$

(where $\bar{p}$ is $F_1(\bar{x}, \bar{x})$) if $(x, y) \in D$ and $(x, y) \neq (\bar{x}, \bar{x})$. Since F is strictly increasing with respect to the first variable, we have $\bar{p} > 0$.
□

Lemma 8.1.8. *Take $\bar{p}$ of Lemma 8.1.7. Then we have:*

$$\forall \varepsilon > 0, \exists \delta^* > 0 \textit{ such that } |x - \bar{x}| \geq \varepsilon \Rightarrow$$

$$F(\bar{x}, \bar{x}) - F(x, y) - \bar{p}y + \bar{p}x > \delta^*, \ \forall y \textit{ with } (x, y) \in D.$$

Proof. Let $E = \{(x, y) \in D : |x - \bar{x}| \geq \varepsilon\}$. Obviously, E is compact. The function ϕ defined by

$$\phi(x, y) = F(\bar{x}, \bar{x}) - F(x, y) - \bar{p}y + \bar{p}x$$

is continuous, strictly positive on E (Lemma 8.1.7) and has a minimum δ^* on E. Thus, we have $\delta^* > 0$. □

Let V_β denote the Value function, i.e.:

$$V_\beta(x_0) = \max \sum_{t=0}^{+\infty} \beta^t F(x_t, x_{t+1}),\ \beta \in [0, 1[,$$

under the constraints:

$$(x_t, x_{t+1}) \in D,\ \forall t \geq 0,\ x_0 \in \mathbb{R}_+ \text{ is given.}$$

One can check, as in the previous chapters, that V_β is continuous, strictly concave. Let g_β denote the optimal policy, i.e.,$\forall x \in [0, a]$,

$$V_\beta(x) = \max\{F(x, y) + \beta V_\beta(y) : (x, y) \in D\} = F(x, g_\beta(x)).$$

Given $x_0 \geq 0$, there is a unique optimal path $(x_0, x_1, ..., x_t, ...)$ and we have $\forall t \geq 1, x_t = g_\beta^t(x_0)$.
In the following lemma, we prove that, given a neighborhood of $\bar{x}$, given the initial capital stock $x_0 > 0$, if β is larger than some value, then the optimal path from x_0 will "visit" this neighborhood at some date t.
Let us recall that a sequence $(x_0, x_1, ..., x_t, ...)$ is feasible from x_0 if $\forall t \geq 0, (x_t, x_{t+1}) \in D$.

Lemma 8.1.9 (The First Visit Lemma). *Let $\underline{x}$ be expansible. Then we have:*

$$\forall \varepsilon > 0, \exists \beta(\varepsilon) > 0, \forall x_0 \geq \underline{x}, \forall \beta \in [\beta(\varepsilon), 1[, \exists \bar{t},\ |g_\beta^{\bar{t}}(x_0) - \bar{x}| < \varepsilon.$$

Proof. Assume the contrary. There exists $\varepsilon > 0$ such that for every β_1, there exists $x_0 \geq \underline{x}, \beta \in [\beta_1, 1[$, and we have $\forall t, |g_\beta^t(x_0) - \bar{x}| \geq \varepsilon$. By Lemma 8.1.8, there exists $\delta^*(\varepsilon) > 0$ such that

$$\forall t, \beta^t F(x_t, x_{t+1}) + \beta^t \bar{p} x_{t+1} - \beta^t \bar{p} x_t < \beta^t F(\bar{x}, \bar{x}) - \beta^t \delta^*(\varepsilon),$$

where $\forall t, x_t = g_\beta^t(x_0)$.
Then,

$$\forall T \geq 0, \sum_{t=0}^{T} \beta^t F(x_t, x_{t+1}) < \sum_{t=0}^{T} \beta^t F(\bar{x}, \bar{x}) - \sum_{t=0}^{T} \beta^t \delta^*(\varepsilon) + \bar{p} x_0.$$

From Lemma 8.1.4, there exists a feasible sequence from $\underline{x}$ which is $(\underline{x}, \underline{x}_1, \underline{x}_2, ..., \underline{x}_N, \bar{x}, \bar{x}, ...)$. From Lemma 8.1.2 (Free disposal), the sequence $(x_0, \underline{x}_1, \underline{x}_2, ..., \underline{x}_N, \bar{x}, \bar{x}, ...)$ is then feasible from x_0. We have

$$\sum_{t=0}^{+\infty} \beta^t F(x_t, x_{t+1}) \geq F(x_0, \underline{x}_1) + ... + \beta^{N-1} F(\underline{x}_{N-1}, \underline{x}_N) + \beta^N F(\underline{x}_N, \bar{x})$$

$$+ \beta^{N+1} \sum_{t=0}^{\infty} \beta^t f(\bar{x}, \bar{x}).$$

Consequently,

$$\sum_{t=0}^{+\infty} \beta^t \delta^*(\varepsilon) \leq -F(x_0, \underline{x}_1) - \beta F(\underline{x}_1, \underline{x}_2) - ... - \beta^{N-1} F(\underline{x}_{N-1}, \underline{x}_N)$$

$$-\beta^N F(\underline{x}_N, \bar{x}) + \sum_{t=0}^{N} \beta^t F(\bar{x}, \bar{x}) + \bar{p} x_0$$

$$\leq NF(\bar{x}, \bar{x}) + \bar{p}a.$$

Thus

$$\frac{1}{1-\beta} \delta^*(\varepsilon) \leq NF(\bar{x}, \bar{x}) + \bar{p}a.$$

This inequality holds for any β_1. Let β_1 converge to 1. Then β converges also to 1. We obtain a contradiction, since the left-hand side of the previous inequality converges to $+\infty$ while the second member remains constant. □

We say that x^β is an optimal steady state, or more simply a steady state, if $x^\beta = g_\beta(x^\beta)$. The optimal sequence from x^β will be the stationary sequence $(x^\beta, x^\beta, ..., x^\beta, ...)$. If (x^β, x^β) is in the interior of D then by the Euler equation, we have $F_2(x^\beta, x^\beta) + \beta F_1(x^\beta, x^\beta) = 0$.

Let us add the following assumption:

H4 The Hessian matrix of F at $(\bar{x}, \bar{x})$ is negative definite.
(For a definition of the Hessian matrix, see Appendix).

Lemma 8.1.10. *For β close to 1, there exists a unique steady state. Moreover, $x^\beta \to \bar{x}$ when $\beta \to 1$.*

Proof. Consider the equation:

$$F_2(x^\beta, x^\beta) + \beta F_1(x^\beta, x^\beta) = 0. \tag{8.1}$$

If we differentiate it, we obtain

$$[F_{21}(x^\beta, x^\beta) + \beta F_{12}(x^\beta, x^\beta) + F_{22}(x^\beta, x^\beta) + \beta F_{11}(x^\beta, x^\beta)]dx^\beta$$
$$+F_1(x^\beta, x^\beta)d\beta = 0,$$

where F_{21}, F_{11}, F_{22} denote the second cross derivatives of F. Let $z = (1,1)$ and H denote the Hessian matrix of F at $(\bar{x}, \bar{x})$. When $\beta = 1$, i.e., when $x^\beta = \bar{x}$, we have, by **H4** :

$$^t zHz = 2F_{21}(\bar{x}, \bar{x}) + F_{22}(\bar{x}, \bar{x}) + F_{11}(\bar{x}, \bar{x}) < 0.$$

Hence, for β close to one, we still have

$$[F_{21}(x^\beta, x^\beta) + \beta F_{12}(x^\beta, x^\beta) + F_{22}(x^\beta, x^\beta) + \beta F_{11}(x^\beta, x^\beta)] < 0.$$

Since for $\beta = 1$ equation (8.1) has a unique solution (Lemma 8.1.6), by the Implicit Functions Theorem (see Appendix), for β close to one, there will be a unique solution x^β which is continuous with respect to β. Hence, $x^\beta \to \bar{x}$ when $\beta \to 1$. □

Lemma 8.1.11 (The Second Visit Lemma). *Let $\underline{x}$ be expansible. Then we have:*
For any $\varepsilon > 0$, there exists $\beta(\varepsilon) > 0$, such that for any $x_0 \geq \underline{x}$, for any $\beta \in [\beta(\varepsilon), 1[$, there exists $\bar{t}$ with $|g_\beta^{\bar{t}}(x_0) - x^\beta| < \varepsilon$.

Proof. From Lemma 8.1.10, there exists β_1 such that, for any $\beta \in [\beta_1, 1[$, we have $|x^\beta - \bar{x}| < \frac{\varepsilon}{2}$. From the First Visit Lemma, we can find β_2 such that, for any $x_0 \geq \underline{x}$, for any $\beta \in [\beta_2, 1[$, there exists $\bar{t}$ with $|g_\beta^{\bar{t}}(x_0) - \bar{x}| < \frac{\varepsilon}{2}$. Take $\beta(\varepsilon) = \max\{\beta_1, \beta_2\}$ to obtain the result. □

8.2 The Neighborhood Turnpike

We now add the following assumption:

H5 There exists $\beta_0 > 0$ such that, if $\beta \in [\beta_0, 1[$, then (x^β, x^β) is in the interior of D. Moreover, for every $x_0 > 0$, for every $\beta \in [\beta_0, 1[$, the optimal path is interior, i.e., $\forall t, (g_\beta^t(x_0), g_\beta^{t+1}(x_0)) \in \text{int } D$.

Throughout the following sections, given x_0, β, we write x_t^β instead of $g_\beta^t(x_0)$.

Lemma 8.2.1. *Let $q^\beta = F_1(x^\beta, x^\beta)$, $q_t^\beta = F_1(x_t^\beta, x_{t+1}^\beta)$. There exists $M > 0$ such that*

$$\forall \beta \in [\beta_0, 1[, \ q^\beta \leq M, \ \forall t, \ q_t^\beta \leq M.$$

Proof. We have $q^\beta = F_1(x^\beta, x^\beta) \leq M_1 = \max\{F_1(x, x) : (x, x) \in P\}$ and $q_t^\beta = F_1(x_t^\beta, x_{t+1}^\beta) \leq M_2 = \max\{F_2(x, y) : (x, y) \in D\}$. Take $M = \max\{M_1, M_2\}$ to end the proof. □

Lemma 8.2.2. *For any $\varepsilon > 0$, there exists $\delta > 0$ such that, for any $\beta \in [\beta_0, 1[$, we have*

$$F(x^\beta, x^\beta) - F(x, y) \geq \beta q^\beta (y - x^\beta) - q^\beta (x - x^\beta) + \delta$$

if (x, y) satisfies : $(x, y) \in D$ and $|x - x^\beta| \geq \varepsilon$.

Proof. Let $\phi(\beta, x, y) = F(x^\beta, x^\beta) - F(x, y) - q^\beta(x^\beta - x) + \beta q^\beta (x^\beta - y)$. Recall that $q^\beta = F_1(x^\beta, x^\beta)$ and by the Euler equation $F_2(x^\beta, x^\beta) = -\beta F_1(x^\beta, x^\beta) = -\beta q^\beta$.
Let

$$\Delta = \{(\beta, x, y) : \beta_0 \leq \beta \leq 1, (x, y) \in D, |x - x^\beta| \geq \varepsilon\}.$$

Obviously, Δ is compact. Since ϕ is continuous, we can find $\delta = \min_{(\beta,x,y)\in\Delta} \phi(\beta, x, y) = \phi(\hat{\beta}, \hat{x}, \hat{y})$. Observe that the function ϕ is non-negative since it is concave and since q^β and $-\beta q^\beta$ are the partial derivatives of F at (x^β, x^β) (see Appendix). But, since F is strictly concave and since $(\hat{x}, \hat{y}) \neq (x^\beta, x^\beta)$, we have $\delta > 0$ (see Appendix). We have proved the lemma. □

Define for $\beta \in [\beta_0, 1[$,

$$L_\beta(t) = (q_t^\beta - q^\beta)(x_t^\beta - x^\beta).$$

Lemma 8.2.3. *We have*
(i) $L_\beta(t) \le 0, \ \forall t, \forall \beta \in [\beta_0, 1[$.
(ii) $\forall t, \forall \beta \in [\beta_0, 1[, \beta L_\beta(t+1) - L_\beta(t) \ge 0$.
(iii) For any $\varepsilon > 0$, *there exists* $\delta > 0$ *such that for every* $x_0 > 0$, *for every* $\beta \in [\beta_0, 1[$, *the following claim is true:*

$$|x_t^\beta - x^\beta| > \varepsilon \Rightarrow \beta L_\beta(t+1) - L_\beta(t) \ge \delta.$$

Proof. (i) Since $(x^\beta, x^\beta) \in \text{int } D, (x_t^\beta, x_{t+1}^\beta) \in \text{int } D, \forall t$, the Value function V_β is differentiable at x^β and at $x_t^\beta, \forall t$. Therefore

$$V_\beta(x_t^\beta) - V_\beta(x^\beta) \ge q_t^\beta(x_t^\beta - x^\beta)$$

$$V_\beta(x_\beta) - V_\beta(x_t^\beta) \ge q^\beta(x^\beta - x_t^\beta).$$

Adding the two inequalities, we obtain $L_\beta(t) \le 0$.
(ii) By the concavity of F, we have

$$F(x^\beta, x^\beta) - F(x_t^\beta, x_{t+1}^\beta) \ge q^\beta(x^\beta - x_t^\beta) - \beta q^\beta(x^\beta - x_{t+1}^\beta).$$

Recall that Euler equation implies

$$F_2(x_t^\beta, x_{t+1}^\beta) + \beta F_1(x_{t+1}^\beta, x_{t+2}^\beta) = 0.$$

We use again the concavity of F to obtain:

$$F(x_t^\beta, x_{t+1}^\beta) - F(x^\beta, x^\beta) \ge q_t^\beta(x_t^\beta - x^\beta) - \beta q_{t+1}^\beta(x_{t+1}^\beta - x^\beta).$$

Adding the two inequalities gives $\beta L_\beta(t+1) - L_\beta(t) \ge 0$.
(iii) From Lemma 8.2.2, if $|x_t^\beta - x^\beta| > \varepsilon$, then

$$F(x^\beta, x^\beta) - F(x_t^\beta, x_{t+1}^\beta) \ge q^\beta(x^\beta - x_t^\beta) - \beta q^\beta(x^\beta - x_{t+1}^\beta) + \delta.$$

But we always have, by the concavity of F:

$$F(x_t^\beta, x_{t+1}^\beta) - F(x^\beta, x^\beta) \ge q_t^\beta(x_t^\beta - x^\beta) - \beta q_{t+1}^\beta(x_{t+1}^\beta - x^\beta).$$

Add the two inequalities to obtain $\beta L_\beta(t+1) - L_\beta(t) \ge \delta$. □

Lemma 8.2.4. *We have the following claim:*
For any $\varepsilon > 0$, *there exists* $L < 0$ *such that, for any* $x_0 > 0$, *for any* $\beta \in [\beta_0, 1[$, *for any* t, *if* $L_\beta(t) \ge L$, *then* $|x_t^\beta - x^\beta| \le \varepsilon$.

Proof. From Lemma 8.2.3(iii), on can find $\delta > 0$ such that

$$|x_t^\beta - x^\beta| > \varepsilon \Rightarrow \beta L_\beta(t+1) \geq \delta + L_\beta(t).$$

Since $L_\beta(t+1) \leq 0$, we will have $L_\beta(t) \leq -\delta$. Hence, if $L_\beta(t) \geq -\frac{\delta}{2}$, we then have $|x_t^\beta - x^\beta| \leq \varepsilon$. □

Theorem 8.2.1 (Neighborhood Turnpike). *Assume* **H1, H2, H3, H4, H5**. *Consider* $\underline{x}$ *of Lemma 8.1.9. Then:*
For any $\varepsilon > 0$*, there exists* $\beta(\varepsilon)$ *such that, for any* $x_0 \geq \underline{x}$*, for any* $\beta \in [\beta(\varepsilon), 1[$*, one can find* T *which satisfies:* $t \geq T \Rightarrow |x_t^\beta - x^\beta| \leq \varepsilon$.

Proof. Let $\varepsilon > 0$. Since $|L_\beta(t)| = |q_t^\beta - q^\beta||x_t^\beta - x^\beta|$, for $\beta \geq \beta_0$, by Lemma 8.2.1, we have $|L_\beta(t)| \leq 2M|x_t^\beta - x^\beta|$. Take L of Lemma 8.2.4 and define $\varepsilon' = \min\{\varepsilon, \frac{\beta_0|L|}{2M}\}$. Take δ' associated with ε' as stated in Lemma 8.2.3(iii), and with $\beta(\varepsilon')$ of Lemma 8.1.11.
Choose $\beta_1 \geq \max\{\beta_0, \beta(\varepsilon')\}$, close to 1 such that $(1-\beta_1)L \geq -\delta'$. By Lemma 8.1.11, there exists T such that $|x_T^{\beta_1} - x^{\beta_1}| \leq \varepsilon' \leq \varepsilon$. Hence, $|L_{\beta_1}(T)| \leq 2M\varepsilon' \leq \beta_0|L|$ or equivalently,

$$0 \geq L_{\beta_1}(T) \geq \beta_0 L \geq \beta_1 L.$$

By Lemma 8.2.3(ii), we have $L_{\beta_1}(T+1) \geq L$ which implies , via Lemma 8.2.4, $|x_{T+1}^{\beta_1} - x^{\beta_1}| \leq \varepsilon$.
If $|x_{T+1}^{\beta_1} - x^{\beta_1}| > \varepsilon'$, then Lemma 8.2.3 (iii) implies:

$$\beta_1 L_{\beta_1}(T+2) - L_{\beta_1}(T+1) \geq \delta'.$$

Therefore

$$\begin{aligned}
L_{\beta_1}(T+2) - L_{\beta_1}(T+1) &\geq \beta_1^{-1}\delta' + (\beta_1^{-1} - 1)L_{\beta_1}(T+1) \\
&= \beta_1^{-1}\delta' + \beta_1^{-1}(1-\beta_1)L_{\beta_1}(T+1) \\
&\geq \beta_1^{-1}\delta' + \beta_1^{-1}(1-\beta_1)L \\
&\geq \beta_1^{-1}\delta' - \beta_1^{-1}\frac{\delta'}{2} = \beta_1^{-1}\frac{\delta'}{2} \geq \frac{\delta'}{2}.
\end{aligned}$$

Hence, $L_{\beta_1}(T+2) > L_{\beta_1}(T+1) \geq L$. Lemma 8.2.4 implies $|x_{T+2}^{\beta_1} - x^{\beta_1}| \leq \varepsilon$.

If $|x^{\beta_1}_{T+2} - x^{\beta_1}| > \varepsilon'$, we repeat the same argument and obtain $|x^{\beta_1}_{T+3} - x^{\beta_1}| \leq \varepsilon$ and $L_{\beta_1}(T+3) - L_{\beta_1}(T+2) \geq \frac{\delta'}{2}$.
If $|x^{\beta_1}_{T+t} - x^{\beta_1}| > \varepsilon'$ for $t = 1, ..., \tau$, we then have:

$$L_{\beta_1}(T+2) - L_{\beta_1}(T+1) \geq \frac{\delta'}{2}, ..., L_{\beta_1}(T+\tau) - L_{\beta_1}(T+\tau-1) \geq \frac{\delta'}{2},$$

and hence,

$$L_{\beta_1}(T+\tau) - L_{\beta_1}(T+1) \geq \frac{\tau-1}{2}\delta'.$$

Then there must be τ such that $|x^{\beta_1}_{T+\tau} - x^{\beta_1}| \leq \varepsilon' \leq \varepsilon$ because, on the contrary, $L_{\beta_1}(T+\tau)$ will become strictly positive in contradiction with Lemma 8.2.3(i). Hence, $|x^{\beta_1}_t - x^{\beta_1}| \leq \varepsilon$ for $t = T, ..., T+\tau$.
If $|x^{\beta_1}_{t+\tau+1} - x^{\beta_1}| > \varepsilon'$, we repeat the argument to find $|x^{\beta_1}_{t+\tau+1} - x^{\beta_1}| \leq \varepsilon$, and so on.
Observe that this result holds for any β_1 larger than some $\beta(\varepsilon)$. In other words, there exists $\beta(\varepsilon)$ such that, for any $\beta \geq \beta(\varepsilon)$, there exists T which satisfies: $t \geq T \Rightarrow |x^{\beta}_t - x^{\beta}| \leq \varepsilon$. □

8.3 Turnpike Theorems

We now come to the main result of this chapter: convergence to a steady state when the discount factor β is close to 1.

Theorem 8.3.1 (First Turnpike Theorem). *Assume* **H1, H2, H3, H4, H5**. *Let $\underline{x}$ be expansible. Then, there exists $\hat{\beta} \in [0, 1[$, such that, for any $x_0 \geq \underline{x}$, for any $\beta \in [\hat{\beta}, 1[$, the optimal sequence $\{g^{\beta}_t(x_0)\}$ converges to x^{β} when $t \to +\infty$.*

Proof. Since F is twice continuously differentiable, the matrix:

$$Q(\beta) = \begin{bmatrix} F_{11}(x^{\beta}, x^{\beta}) & F_{12}(x^{\beta}, x^{\beta}) \\ \frac{1}{\beta}F_{21}(x^{\beta}, x^{\beta}) & \frac{1}{\beta}F_{22}(x^{\beta}, x^{\beta}) \end{bmatrix}$$

is negative definite (for a definition, see Appendix) for β close enough to 1, say, $\beta \geq \beta_1$. For any $(x, y) \in \mathbb{R}^2$, we have

$$\begin{pmatrix} x, & y \end{pmatrix} Q(\beta) \begin{pmatrix} x \\ y \end{pmatrix} = \frac{1}{2} \begin{pmatrix} x, & y \end{pmatrix} (Q(\beta) + Q'(\beta)) \begin{pmatrix} x \\ y \end{pmatrix}$$

where $Q'(\beta)$ denotes the transpose matrix of $Q(\beta)$. Since $(Q(\beta) + Q'(\beta))$ is symmetric, negative definite, we have:

$$\begin{pmatrix} x, & y \end{pmatrix} (Q(\beta) + Q'(\beta)) \begin{pmatrix} x \\ y \end{pmatrix} \leq \lambda(\beta)(x^2 + y^2)$$

where $\lambda(\beta)$ is the negative eigenvalue of the matrix $\frac{1}{2}(Q(\beta)+Q'(\beta))$ which has the smallest modulus (see Appendix).
Let us recall that

$$L_\beta(t-1) = (F_1(x_{t-1}^\beta, x_t^\beta) - F_1(x^\beta, x^\beta))(x_{t-1}^\beta - x_{t-1}),$$

$$L_\beta(t) = (F_1(x_t^\beta, x_{t+1}^\beta) - F_1(x^\beta, x^\beta))(x_t^\beta - x_t),$$

and, by the Euler equation:

$$L_\beta(t) = -\frac{1}{\beta}(F_2(x_{t-1}^\beta, x_t^\beta) - F_2(x^\beta, x^\beta))(x_t^\beta - x_t).$$

By the first order expansion, we have:

$$F_1(x_{t-1}^\beta, x_t^\beta) - F_1(x^\beta, x^\beta) =$$

$$F_{11}(x^\beta, x^\beta)(x_{t-1}^\beta - x^\beta) + F_{12}(x^\beta, x^\beta)(x_t^\beta - x^\beta) + o_1(x^\beta, x_t^\beta, x_{t-1}^\beta)$$

and

$$F_2(x_{t-1}^\beta, x_t^\beta) - F_2(x^\beta, x^\beta) =$$

$$F_{21}(x^\beta, x^\beta)(x_{t-1}^\beta - x^\beta) + F_{22}(x^\beta, x^\beta)(x_t^\beta - x^\beta) + o_2(x^\beta, x_t^\beta, x_{t-1}^\beta).$$

Therefore

$$\begin{aligned} L_\beta(t) - L_\beta(t-1) = &- \begin{pmatrix} x_{t-1}^\beta - x^\beta, & x_t^\beta - x^\beta \end{pmatrix} Q(\beta) \begin{pmatrix} x_{t-1}^\beta - x^\beta \\ x_t^\beta - x^\beta \end{pmatrix} \\ &-\frac{1}{\beta} o_2(x^\beta, x_t^\beta, x_{t-1}^\beta)(x^\beta - x^\beta) \\ &- o_1(x^\beta, x_t^\beta, x_{t-1}^\beta)(x_{t-1}^\beta - x^\beta) \\ \geq & \left[\frac{1}{2}|\lambda(\beta)| - \epsilon(x^\beta, x_t^\beta, x_{t-1}^\beta)\right] \left\|(x_{t-1}^\beta - x^\beta, x_t^\beta - x^\beta)\right\|^2, \end{aligned}$$

where

$$\epsilon(x^\beta, x_t^\beta, x_{t-1}^\beta) = \frac{|o_2(x^\beta, x_t^\beta, x_{t-1}^\beta)| + |o_1(x^\beta, x_t^\beta, x_{t-1}^\beta)|}{\beta \|(x_{t-1}^\beta - x^\beta, x_t^\beta - x^\beta)\|},$$

and

$$\|(x_{t-1}^\beta - x^\beta, x_t^\beta - x^\beta)\| = (|x_{t-1}^\beta - x^\beta|^2 + |x_t^\beta - x^\beta|^2)^{\frac{1}{2}}.$$

There exists $\varepsilon > 0$ such that, if $\|(x_{t-1}^\beta - x^\beta, x_t^\beta - x^\beta)\| \leq \varepsilon$, then

$$\left[\frac{1}{2}|\lambda(\beta)| - \epsilon(x^\beta, x_t^\beta, x_{t-1}^\beta)\right] \geq \frac{1}{4}|\lambda(\beta)|.$$

Choose such an ε. From Theorem 8.2.1, there exists β_2 (we can choose $\beta_2 \geq \beta_1$) such that, for any $\beta \in [\beta_2, 1[$, there exists T which satisfies $\forall t \geq T, |x_t^\beta - x^\beta| \leq \varepsilon$. Take $\beta \in [\beta_2, 1[$ and the associated T. Let $s_\beta(t) = L_\beta(t) - L_\beta(t-1)$. For every $t \geq T+1$, we have

$$s_\beta(t) \geq \frac{1}{4}|\lambda(\beta)| \|(x_{t-1}^\beta - x^\beta, x_t^\beta - x^\beta)\|^2 \geq 0. \tag{8.2}$$

We also have:

$$\forall \tau \geq 1, \sum_{t=T+1}^{T+\tau} s_\beta(t) = L_\beta(T+\tau) - L_\beta(T) \leq -L_\beta(T),$$

since $L_\beta(t) \leq 0, \forall t$ (see Lemma 8.2.3). Hence, $\sum_{T+1}^{+\infty} s_\beta(t) < +\infty$ and $s_\beta(t)$ converges to 0 when t converges to $+\infty$. From inequality (8.2), we conclude that $x_t^\beta \to x^\beta$ when $t \to +\infty$. □

We will now prove the turnpike for unbounded technology sets. We maintain assumption **H2** but replace **H1** by
H'1 The technology set D is

$$D = \{(x, y) \in \mathbb{R}_+^2 : y \leq f(x)\}$$

with $a > 0$, and f, a differentiable, strictly concave, increasing function. The function f satisfies: $f(0) = 0$, there exists $\hat{x} > 0$ with $f(\hat{x}) = \hat{x}$.

We then obtain the second turnpike theorem.

Theorem 8.3.2 (Second Turnpike Theorem). *Assume* **H'1, H2, H3, H4, H5**. *Let* $\underline{x}$ *be expansible. Then, there exists* $\hat{\beta} \in [0,1[$, *such that, for any* $x_0 \geq \underline{x}$, *for any* $\beta \in [\hat{\beta}, 1[$, *the optimal sequence* $\{g_t^\beta(x_0)\}$ *converges to* x^β *when* $t \to +\infty$.

Proof. The reader can check that Lemma 8.1.1 still holds. Consider the number R in the statement (ii) of this lemma. Observe that $R > \hat{x}$. From Lemma 8.1.3, any expansible stock is smaller than R. Let $\underline{x}$ be expansible. From the First Turnpike Theorem, there exists $\hat{\beta} \in [0,1[$, such that, for any x_0 which satisfies $\underline{x} \leq x_0 \leq R$, for any $\beta \in [\hat{\beta}, 1[$, the optimal sequence $\{g_t^\beta(x_0)\}$ converges to x^β when $t \to +\infty$.
Assume now $x_0 > R$. Let $\mathbf{x}$ be the optimal path from x_0. From statement (ii) of Lemma 8.1.1, there exists T such that $x_T \leq R$. From the optimality principle, the sequence $(x_T, x_{T+1}, ..., x_{T+t}, ...)$ is optimal for an initial stock equal to x_T. From the First Turnpike Theorem, it converges to x^β. In other words, the whole sequence $\mathbf{x}$ converges to x^β. □

8.4 Remarks

1. The proofs of the Visit Lemmas are based Scheinkman's paper [76] while those of the Neighborhood Turnpike and Turnpike are borrowed from McKenzie's paper [56].
2. Scheinkman [76] used the continuity properties of the stable manifolds associated with the Euler equation to prove the local convergence to the steady state of the optimal path. His proof of the global convergence is done in three steps.
Step 1. He proved the Visit Lemmas.
Step 2. Consider the Euler equation for the optimal path (x_t^β):

$$F_2(x_t, x_{t+1}) + \beta F_1(x_{t+1}, x_{t+2}) = 0, \ \forall t \geq 0.$$

Recall that

$$F_2(x_t^\beta, x_{t+1}^\beta) + \beta F_1(x_{t+1}^\beta, x_{t+2}^\beta) = 0,$$

and

$$F_2(\bar{x}, \bar{x}) + F_1(\bar{x}, \bar{x}) = 0.$$

Scheinkman assumed that $F_{12}(\bar{x}, \bar{x}) \neq 0$ (if we have an n-sector model, then the matrix $F_{12}(\bar{x}, \bar{x})$ is assumed to be invertible). In this case, for β sufficiently near 1, there exists a neighborhood $\mathcal{V}$ of $(\bar{x}, \bar{x}, \bar{x})$ such that the system

$$F_2(x, y) + \beta F_1(y, z) = 0$$

with $(x, y, z) \in \mathcal{V}$, is equivalent to

$$z = \gamma_\beta(x, y)$$

where γ_β is twice continuously differentiable in $\mathcal{V}$.
If we discretize the Euler equation corresponding to $\beta = 1$

$$F_2(x_t, x_{t+1}) + F_1(x_{t+1}, x_{t+2}) = 0$$

around $(\bar{x}, \bar{x}, \bar{x})$, we obtain the system:

$$F_{21}(\bar{x}, \bar{x})X_t + (F_{22}(\bar{x}, \bar{x}) + F_{11}(\bar{x}, \bar{x}))X_{t+1} + F_{12}(\bar{x}, \bar{x})X_{t+2} = 0.$$

One can check that, in our one-sector model, this system possesses two ($2n$ if we have n sectors) characteristic roots, one of them has a modulus larger than one, the other one has a modulus smaller than one (for a n- sector model, we will have n characteristic roots with modulus larger than one, and n roots with modulus smaller than one). We say that $\bar{x}$ possesses the *saddle-point* property. Scheinkman [76] shows that this property implies the existence of $\beta_0 > 0, \varepsilon > 0$ such that, if $\beta \in [\beta_0, 1]$ and if $|x_0 - x^\beta| \leq \varepsilon$, then there exists x_1 with $|x_1 - x^\beta| \leq \varepsilon$ for which the sequence $(\gamma_\beta(x_t, x_{t+1}))_{t=0,1,..,+\infty}$ converges to x_β when t converges to $+\infty$.
Step 3. Let (x_t^β) be the optimal path (we assume that it is interior and hence, satisfies the Euler equation). From Step 1, there exists $\beta_1(\varepsilon)$ such that, for any $\beta \in [\beta_1(\varepsilon), 1[$, there is T with $|x_T^\beta - x^\beta| \leq \varepsilon$. Let $y_0 = x_T^\beta$. From Step 2, there exists y_1 for which the path $(\gamma_\beta(y_t, y_{t+1}))$ converges to x^β. Define the sequence $\mathbf{x}'$ as follows

$$x'_t = x_t, \ \forall t = 1, ..., T-1 \text{ and } x'_{T+t} = y_t, \ \forall t = 0, 1, ..., +\infty.$$

The sequence $\mathbf{x}'$ satisfies the Euler equation and the Transversality condition (since it converges to x^β). From Mangasarian Lemma (see Chapter 4), it is therefore optimal . Since the optimal path is unique, it coincides with $\mathbf{x}^\beta$. Take $\beta(\varepsilon) = \max\{\beta_0, \beta_1(\varepsilon)\}$ to obtain the Turnpike theorem (Theorem 8.3.1).

9. Cycles and Chaos in Optimal Growth Models

In this last chapter, we discuss the occurrence of optimal cycles and optimal chaos in optimal growth models. As most of the results in the literature deal with models in which the dynamics can be reduced to a one dimensional difference equation, we will restrict ourselves to that case. In particular, we devote more than half of the chapter to the properties of the neoclassical two-sector growth model. A section deals with optimal chaos. Although we introduce a few concepts from the theory of dynamical systems, we have chosen to emphazise the " everything goes approach" to optimal chaos.

9.1 Two-Sector Models

9.1.1 The model

We first recall the two-sector optimal growth model that was introduced in chapter4. There are two goods: (i) a consumption good c produced by sector 1 through a production function $f^c(k^1, l^1)$ using capital stock k^1 and a quantity of labor l^1, (ii) a capital good k produced by sector 2 through a production function f^k using k^2 capital and l^2 labor. We assume that the supply of labor is constant over time and equal to 1. The functions f^c and f^k are concave, homogeneous of degree one, strictly increasing, twice continuously differentiable almost every where and satisfy $f^c(0,l) = f^c(k,0) = f^k(0,l) = f^k(k,0) = 0$ for all $k \geq 0$, for all $l \geq 0$.
Assuming to simplify that the consumer's utility is linear ($u(c) = c$) and that the rate of depreciation is zero, the consumer solves the problem:

$$\max \sum_{t=0}^{+\infty} \beta^t u(c_t), \text{ with } 0 < \beta < 1,$$

under the constraints:

$$\forall t \geq 0,\ 0 \leq c_t \leq f^c(k_t^1, l_t^1),$$
$$0 \leq k_{t+1} \leq f^k(k_t^2, l_t^2),$$
$$k_t^1 + k_t^2 \leq k_t,$$
$$l_t^1 + l_t^2 \leq 1,$$
$$l_t^1 \geq 0,\ l_t^2 \geq 0,\ k_t^1 \geq 0,\ k_t^2 \geq 0,$$

and $k_0 > 0$ is given.
Let the correspondence $\Gamma : \mathbb{R}_+ \to \mathbb{R}_+$ be defined by $\Gamma(k) = [0, f^k(k,1)]$ and $F : \text{graph}(\Gamma) \to \mathbb{R}_+$ be defined as follows:

$$F(k,y) = \max_{k^1,l^1,k^2,l^2} f^c(k^1, l^1),$$

under the constraints

$$0 \leq k^1 + k^2 \leq k,$$
$$0 \leq l^1 + l^2 \leq 1,$$
$$0 \leq y \leq f^k(k^2, l^2),$$
$$k^1 \geq 0,\ k^2 \geq 0,\ l^1 \geq 0,\ l^2 \geq 0.$$

Let $((k^i(k,y), l^i(k,y)),\ i = 1,2)$ denote the optimal capital-labor pairs associated with a given input-output pair (k,y). The initial problem is equivalent to the following :

$$\max \sum_{t=0}^{+\infty} \beta^t F(k_t, k_{t+1}),$$

under the constraints

$$0 \leq k_{t+1} \in \Gamma(k_t),\ \forall t \geq 0,$$

and $k_0 > 0$ is given. Clearly the assumptions **H1**, **H'2**, **H'3** of Chapter 4 and **H2** of Chapter 8 are verified.

Proposition 9.1.1. *Assume that F is strictly concave and C^2 on the interior of graph(Γ).*
Then $F_{12}(k,y) > 0$ (resp $F_{12}(k,y) < 0$) iff $\frac{k^2(k,y)}{l^2(k,y)} > \frac{k^1(k,y)}{l^1(k,y)}$ $\left(\text{resp } \frac{k^1(k,y)}{l^1(k,y)} > \frac{k^2(k,y)}{l^2(k,y)}\right)$. In other words $F_{12}(k,y) > 0$ iff the factor intensity of the capital good sector is smaller than the factor intensity of the consumption good. Therefore if the optimal policy is interior, it is increasing (resp decreasing) iff $\frac{k^1(k,y)}{l^1(k,y)} < \frac{k^2(k,y)}{l^2(k,y)}$ $\left(\text{resp } \frac{k^1(k,y)}{l^1(k,y)} > \frac{k^2(k,y)}{l^2(k,y)}\right)$.

Proof. The proof of the first assertion of the proposition may be found in the appendix of this chapter. The second assertion follows directly from proposition 4.2.4. □

9.1.2 Examples

For further use, let us consider a few examples.

Example 1

Assume that $f^c(k,1)$ is C^2 for $k > 0$ and that the production function of the capital sector is of Leontieff type :$f^k(k,l) = \min\{l, \frac{k}{a}\}$, with $0 < a < 1$. Let $\Gamma(k) = [0, \min\{1, \frac{k}{a}\}]$ and $F(k,y) = f^c(k - ay, 1 - y)$. F is C^2 on the interior of graph(Γ). Using the homogeneity of f^c, one gets

$$F_{12}(k,y) = \frac{1}{l^1(k,y)} \frac{\partial^2 f_c}{\partial^2 k} \left(\frac{k - ay}{1 - y}, 1\right) \left[\frac{k^1(k,y)}{l^1(k,y)} - \frac{k^2(k,y)}{l^2(k,y)}\right]$$

with $l^2(k,y) = y$, $k^2(k,y) = ay$. Hence the result of the proposition holds true when the capital sector has a Leontieff technology.

Assume further that $f^c(k,l) = k^{1-\alpha} l^\alpha$, with $0 < \alpha < 1$. Then

$$F(k,y) = (k - ay)^{1-\alpha}(1 - y)^\alpha.$$

This example was studied by Boldrin and Deneckere [16].

Example 2

Let us next assume that both technologies are given by Cobb-Douglas functions: $f_c(k,l) = k^{1-\alpha}l^{\alpha}$ and $f_k(k,l) = k^{1-\gamma}l^{\gamma}$, with $0 < \alpha < 1$ and $0 < \gamma < 1$. Then $\Gamma(k) = [0, k^{1-\gamma}]$ and it follows from the first order conditions that

$$\frac{\frac{\partial f^c}{\partial k}(k^1, l^1)}{\frac{\partial f^k}{\partial k}(k^2, l^2)} = \frac{\frac{\partial f^c}{\partial l}(k^1, l^1)}{\frac{\partial f^k}{\partial l}(k^2, l^2)}$$

Therefore

$$\frac{\frac{k^2(k,y)}{l^2(k,y)}}{\frac{k^1(k,y)}{l^1(k,y)}} = \frac{\frac{1-\alpha}{\alpha}}{\frac{\gamma}{1-\gamma}}$$

and $\frac{k^2(k,y)}{l^2(k,y)} > \frac{k^1(k,y)}{l^1(k,y)}$ iff $\gamma > \alpha$.

We shall later give the precise expression of $F(k,y)$ in the special case $\alpha = \frac{1}{2}$.

Example 3

We present this example because of its historic interest although it doesn't fulfill the hypothesis $f^k(0,l) = 0$ for all $l \geq 0$. Assume that $f^c(k,l) = k^{1-\alpha}l^{\alpha}$ and $f^k(k,l) = l$. It may easily be checked that $\Gamma(k) = [0,1]$ and

$$F(k,y) = k^{\alpha}(1-y)^{1-\alpha}$$

Note that $\Gamma(0) \neq \{0\}$. This indirect utility was first studied by Samuelson [72] in the case $\alpha = \frac{1}{2}$, then by Mckenzie [55] and Benhabib and Nishimura [11] as an example of utility having an optimal policy with a period two-cycle.

9.1.3 Optimal interior stationary state

The Euler's equation of the optimization problem is as usual

$$F_2(k_{t-1}, k_t) + \beta F_1(k_t, k_{t+1}) = 0.$$

An interior steady state $k(\beta)$ corresponds to a stationary solution of Euler's equation:

$$F_2(k(\beta), k(\beta)) + \beta F_1(k(\beta), k(\beta)) = 0.$$

Let us remark that for β close to 1, there is a unique stationary solution. Indeed, as mentioned in chapters 7 and 8, there exists a unique $\bar{k}$ which maximizes $F(k,k)$ for $k \in \Gamma(k)$. Whenever interior, $\bar{k}$ is the unique solution to the equation

$$F_2(k,k) + F_1(k,k) = 0$$

Therefore

$$F_2(k,k) + \beta F_1(k,k) = 0$$

has a unique solution for $\beta = 1$. By the Implicit functions theorem (see the Appendix), it also has a unique solution for β close to 1. For smaller β, there may be several interior steady states.

The local behavior around a stationary point $k(\beta)$ is determined by the linearized equation at $k(\beta)$

$$\beta F_{12}(\beta) k_{t+1} + [\beta F_{11}(\beta) + F_{22}(\beta)] k_t + F_{12}(\beta) k_{t-1} = 0.$$

with $F_{ij}(\beta) = F_{ji}(k(\beta), k(\beta)), i = 1,2, j = 1,2$. The characteristic roots are solutions to the following equation:

$$\beta F_{12}(\beta)\lambda^2 + [\beta F_{11}(\beta) + F_{22}(\beta)]\lambda + F_{21}(\beta) = 0$$

Proposition 9.1.2. – *The product of the roots is* $\frac{1}{\beta}$.

– *If the Hessian matrix of* F *at* $(k(\beta), k(\beta))$ *is definite negative, then for any root* λ, $|\lambda| \neq (\frac{1}{\beta})^{\frac{1}{2}}$. *The roots are real and some root* λ *has modulus smaller than* $(\frac{1}{\beta})^{\frac{1}{2}}$. *If furthermore* $F_{12}(\beta) > 0$ *(resp* $F_{12}(\beta) < 0$*), then both roots are positive (resp negative).*

– *If* F *is* C^2 *and for some* $L > 0$, $F(k,y) + \frac{L}{2}y^2$ *is concave, then the policy function is differentiable and* $g'(k(\beta)) = \lambda$ *with* $|\lambda| < (\frac{1}{\beta})^{\frac{1}{2}}$.

Proof. The first assertion is obvious from the characteristic equation. To prove the second, assume that the roots are complex. Then $|\lambda^2| = \frac{1}{\beta}$. Multiplying the characteristic equation by $\bar{\lambda}$ and using the equality $\lambda\bar{\lambda} = \frac{1}{\beta}$, we get

$$\beta F_{12}(\beta)\lambda + [\beta F_{11}(\beta) + F_{22}(\beta)]\frac{1}{\beta} + F_{21}(\beta)\bar{\lambda} = 0$$

Equivalently,

$$[1 \quad \bar{\lambda}] \begin{bmatrix} F_{11}(\beta) & F_{21}(\beta) \\ F_{21}(\beta) & F_{22}(\beta) \end{bmatrix} \begin{bmatrix} 1 \\ \lambda \end{bmatrix} = 0$$

contradicting the assumption that the determinant of the Hessian matrix of F at $(k(\beta), k(\beta))$ is different from zero. Hence the characteristic roots are real and $|\lambda| \neq (\frac{1}{\beta})^{\frac{1}{2}}$.
To prove the last statement of the second assertion, observe that the sum of the roots is equal to $-\frac{\beta F_{11}(\beta)+F_{22}(\beta)}{\beta F_{12}(\beta)}$ which is strictly positive if $F_{12}(\beta) > 0$.
The proof of the last assertion is more difficult and may be found in Santos [74]. □

Remark 9.1.1. The assumption that for some $L > 0$, $F(k, y) + \frac{L}{2}y^2$ is concave is known as L_k concavity of F. It was introduced in the growth literature by Boldrin and Montrucchio [17].

9.1.4 Difficulties

Since the product of the characteristic roots is equal to $\frac{1}{\beta} > 1$, at least one root has modulus greater than one. We have seen in Chapter 8 that when β is close to one, one root of the characteristic equation has modulus smaller than one. Hence we have local convergence in a neighborhood of $k(\beta)$. What happens when $k(\beta)$ looses its stability?

The theory of one dimensional non linear dynamics is precisely concerned with the existence of periodic points and chaotic behavior for a map of an interval to itself. Changes in the properties of periodic points and in the asymptotic behavior of a family of maps of an interval to itself that depends on a parameter are called bifurcations. It turns out that even when the family of maps is explicitly known and a number of regularity assumptions are fulfilled, the study of its bifurcations is not easy (the reader is referred to the survey article of Grandmont [40]). In optimal growth problems, the discount factor appears as a natural parameter. In most optimal growth models, the optimal policy is not known explicitly. The use of bifurcation theory is in general inadequate because it requires conditions on the third derivative of the policy function. Hence, we shall only use here topological proofs to prove existence of periodic orbits .

9.1.5 Existence of periodic orbits.

To simplify the presentation, let us assume that the optimal policy maps $[0,1]$ into itself. Let us recall that a continuous map $g : [0,1] \to [0,1]$ is unimodal if there exists $\bar{k}$ such that g is increasing on $[0,\bar{k}]$ and decreasing on $[\bar{k},1]$.

Proposition 9.1.3. *1. Assume that the optimal policy g is unimodal, that there exists a unique interior unstable steady state $k(\beta)$ and that g is differentiable at $k(\beta)$ with $g'(k(\beta)) > 1$. Then there exists a period two orbit.*

2. If $F_{12} < 0$ on the interior of graph(Γ) and if $f^k(k,1) > k$ on $[0,1]$, then the optimal policy g is unimodal and there exists a unique interior steady state.

Proof. The existence of a period two orbit for g is equivalent to the existence of a fixed point for g^2. If the optimal policy g is unimodal and there exists a unique interior unstable steady state $k(\beta)$, then $g(k) > k$ on $[0,\bar{k}]$. Let $M = g(\bar{k})$. Since g is increasing on $[0,\bar{k}]$ and decreasing on $[\bar{k},1]$, $M = \max_{[0,1]} g(k)$. Let $k_1 \in [0,\bar{k}]$ be such that $g(k_1) = \bar{k}$. On $[0,k_1]$, g^2 is increasing and furthermore $g^2(k_1) = g(\bar{k}) > \bar{k} > k_1$. Since $(g^2)'(k(\beta)) > 1$, there exists $k_2 \in]\bar{k}, k(\beta)[$ such that $g^2(k_2) < k_2$. Since g^2 is continuous, it has a fixed point in $[\bar{k}, k_2]$ which proves the first assertion.
To prove the second assertion, assume that $F_{12} < 0$ on the interior of graph(Γ). Since $\Gamma(k) = [0, f^k(k,1)]$ and $f^k(0,1) = 0$, $\Gamma(0) = \{0\}$, $g(0) = 0$. Hence g is non decreasing in a neighborhood of 0. The optimal policy g cannot be interior in that neighborhood since if it were as $F_{12} < 0$, it would be decreasing. It cannot be zero on an interval $[0,a]$ since it would be increasing in a neighborhood of a contradicting the assumption that $F_{12} < 0$. Hence $g(k) = f^k(k,1)$

on some interval $[0, \bar{k}]$. This implies that g is concave, increasing on that interval and above the diagonal since $f^k(k, 1) > k$. On $[\bar{k}, 1]$, g being interior is decreasing, hence g is unimodal. It has a unique optimal interior steady state $k(\beta) \in [\bar{k}, 1]$. □

Remark 9.1.2. We have not proven any result about the stability of the period two orbit.

An example.

This example is due to Nishimura and Yano [63]. As in Example 2, let us assume that both technologies are given by Cobb-Douglas functions: $f^c(k, l) = k^{1-\alpha} l^{\alpha}$ and $f^k(k, l) = k^{\frac{1}{2}} l^{\frac{1}{2}}$, with $\frac{1}{2} < \alpha < 1$. It follows that $\frac{k^2(k,y)}{l^2(k,y)} < \frac{k^1(k,y)}{l^1(k,y)}$, hence that $F_{12} < 0$ on interior of graph Γ where $\Gamma(k) = [0, \sqrt{k}]$. We have

$$F(k, y) = \left(\frac{1-\alpha}{(1-\alpha)k + (2\alpha - 1)h(k, y)} \right)^{1-\alpha} [k - h(k, y)]$$

where

$$h(k, y) = \frac{(2\alpha - 1)y^2 + \sqrt{(2\alpha - 1)^2 y^4 + 4\alpha(1-\alpha)y^2 k}}{2\alpha}$$

There is a unique stationary state

$$k(\beta) = \frac{\alpha\beta^2}{2[2(1-\alpha) + \beta(2\alpha - 1)]}$$

The characteristic equation has two negative roots:

$$\lambda = -\frac{2(1-\alpha)}{\beta(2\alpha - 1)} \quad \text{and} \quad \lambda' = -\frac{2\alpha - 1}{2(1-\alpha)}$$

If $\frac{3}{4} < \alpha < 1$ and $\beta < \frac{2(1-\alpha)}{2\alpha - 1}$, then both roots have modulus larger than one, hence $k(\beta)$ is unstable. Therefore if $\frac{3}{4} < \alpha < 1$, there is a period two optimal path for any $\beta < \frac{2(1-\alpha)}{2\alpha-1}$.

9.2 Optimal chaos

The existence of chaotic optimal policies can been proven in two ways. One can show the occurrence of complicated dynamics for the optimal solution to a given optimal growth model. This approach obviously requires a definition of what do we mean by "complicated dynamics". In the spirit of the "everything goes" market excess demand characterization of Debreu [28], Mantel [53], Sonnenschein [78], one can show that any map from an interval into itself is the solution to an optimal growth model. This is what we call the "rationalizability" of a map. We devote the next two subsections to these two approaches.

9.2.1 Two definitions of complicated dynamics

Ergodic chaos.
For this section, the reader can refer to Stokey and Lucas [65] for the definitions of the Borel sigma- field, measures, and measurable functions.

We give two definitions of complicated dynamics.

Let $g : [0,1] \to [0,1]$ be a continuous map and let $x_{t+1} = g(x_t)$ be the associated dynamical system. Let $\mathcal{B}$ denote the Borel sigma-field of $[0,1]$.

Definition 9.2.1. *A probability measure μ on $([0,1], \mathcal{B})$ is absolutely continuous with respect to Lebesgue measure λ if $\lambda(A) = 0$ implies $\mu(A) = 0$. It is invariant with respect to g if $\mu(g^{-1}(B)) = \mu(B)$ for every Borel set of $[0,1]$. It is ergodic if:*

$$\frac{1}{n}\sum_{j=1}^{n} h(g^{j-1}(x)) \to \int_{[0,1]} h d\mu$$

for any μ measurable function h and for μ almost every $x \in [0,1]$. In other words, time averages along a trajectory can be replaced by space averages.

This implies in particular that for almost any x, the empirical distribution $\nu_n(x)$ that is generated by the iterates $g^j(x),\ j = 0, \ldots, n-1$

which assign probability $\frac{1}{n}$ to each $g^j(x)$, converges to ν weakly. In other words, if μ is absolutely continuous with respect to Lebesgue measure and ergodic, the iterates of x eventually fill up the support of the measure μ. A sufficient condition for existence of an absolutely continuous invariant measure is given in Lasota and Yorke [46].

Definition 9.2.2. *A map $g : [0,1] \to [0,1]$ is piecewise expanding if there exists a finite set of points $a_0 < a_1 < a_2 < \ldots < a_m = 1$ such that on each interval $[a_{i-1}, a_i]$, g is C^1 and $|g'(x)| \geq A > 1$*

Lemma 9.2.1. *If g is piecewise expanding and the restriction of g to each interval $[a_{i-1}, a_i]$ is C^2 and on-to $[0,1]$, then there exists an absolutely continuous invariant measure for g*

Proof. See Lasota and Yorke [46]. □

Example.
A map of the interval that fulfills definition 9.2.2 is the tent map $g(x) = 1 - |2x - 1|$. It has been extensively studied in the one dimensional non linear dynamics literature. Sorger [79] shows that the tent map is the optimal solution to the growth problem (F, Γ, β) with

$$F(k,y) = (2 - \frac{1}{2\beta})k^2 - 4k + (2\beta - \frac{1}{2})y^2 + 2y$$

$\Gamma(k,y) = \{(k,y) \mid 0 \leq k \leq 1,\ 0 \leq y \leq 2k,\ 0 \leq y \leq 2(1-k)\}$ for discount factors β smaller than $\frac{1}{4}$. The value function is $W(k) = -(2 + \frac{1}{2\beta})\, k^2$.

Topological chaos.
Let $g : [0,1] \to [0,1]$ be a continuous map and let $x_{t+1} = g(x_t)$ be the associated dynamical system. Let $p \geq 2$. We recall that $\bar{x}$ is a periodic point of order p if $g^p(\bar{x}) = \bar{x}$. The corresponding cycle $(\bar{x}, x_2, \ldots, x_{k-1}, \bar{x})$ is the set of iterates $x_j = g^j(\bar{x})$ of $\bar{x}$. The periodic point is stable if there exists a neighborhood U of $\bar{x}$ such that for every x in U, $g^{kj}(x) \in U$ and $\lim_{j\to\infty} g^{kj}(x) = \bar{x}$. If g is differentiable at $\bar{x}$, the periodic point is stable iff $(g^k)'(\bar{x}) < 1$.

Definition 9.2.3. *The map g exhibits topological chaos if :*
a) there are infinitely many periodic points with different periods and
b) there is an uncountable set $S \subseteq [0,1]$ *containing no periodic points such that*

$$\liminf_{t\to\infty} |g^t(x) - g^t(y)| = 0 < \limsup_{t\to\infty} |g^t(x) - g^t(y)|$$

for every $x \in S, y \in S$ *and* $x \neq y$ *and*

$$\limsup_{t\to\infty} |g^t(x) - g^t(y)| > 0$$

for every $x \in S$ *and* y *periodic.*

The set S is called a scrambled set. The interpretation of the condition stated above is that any two trajectories starting in the scrambled set move arbitrarily close to each other but do not converge to each other or to a periodic point. Although the scrambled set is required to be uncountable, it can be a set of zero Lebesgue measure. Thus the name "topological chaos".

Let us state a lemma which is due to Li and Yorke [49] and Sarkovski [75].

Lemma 9.2.2. *If g has a periodic point of order* 3*, then g is chaotic and it has a periodic point of every order. More generally if g has a periodic point which is not a power of two, then g is chaotic.*

Proof. Refer to Li and Yorke [49] and Sarkovski [75]. □

Generalizing a result of Sorger [80], Mitra [58] and Nishimura and Yano [64] show that if $F(k,.)$ is strictly concave for every K and if I is an interval of $\mathbb{R}$ and $g : I \to I$ exhibits Li and Yorke chaos, then $\beta < \left(\frac{\sqrt{5}-1}{4}\right)^2 = 0,382$.

9.2.2 "Rationalizability" of a map

We have seen in Proposition 4.2.4. that under **H1**, **H'2**,**H'3** of Chapter 4 and **H2** of Chapter 8, the optimal solution of an optimal growth problem is continuous. This raises the question whether any continuous map is an optimal policy. We state a theorem for a map $g : [0,1] \to [0,1]$ but we may more generally consider a map defined on a compact state space of arbitrary finite dimension.

Boldrin-Montrucchio-Sorger's construction.

Theorem 9.2.1. *Let $g : [0,1] \to [0,1]$ be a twice continuously differentiable function. Let Γ be a correspondence fulfilling* **H1**, **H'2** *of Chapter 4 such that $g(k) \in \Gamma(k)$ for all $k \in [0,1]$. Then there exists a number β^* such that for all $\beta \in (0, \beta^*)$, there exists a utility function F that fulfills* **H'3** *of Chapter 4,* **H2** *of Chapter 8 having g as optimal policy.*

Proof. We provide a proof due to Sorger [79]. It simplifies the original proof of Boldrin and Montrucchio [17]. Let

$$F(k,y) = -\frac{1}{2}[y - g(k)]^2 - \frac{a}{2}k^2 + \frac{a\beta}{2}y^2 + bk - by\beta$$

where a and b are real parameters with $a > 0$. Let $W(k) = -\frac{a}{2}k^2 + bk$. Since

$$\begin{aligned} W(k) &= F(k, g(k)) + \beta W(g(k)) \\ &\geq F(k,y) + \beta W(y) = W(k) - \frac{1}{2}[y - g(k)]^2, \end{aligned}$$

for all $y \in [0,1]$ and $g(k) \in \Gamma(k)$ for all $k \in [0,1]$, g is the optimal solution to the optimal growth problem associated to (F, Γ, β). It remains to prove that one can choose a and b so that F fulfills **H'3** of Chapter 4, **H2** of Chapter 8. The sum S and product P of the characteristic roots of the Hessian of F at (k,y), $H_F(k,y)$ are respectively $S = -g'(k)^2 + (y - g(k))g''(k) - a - 1 + a\beta$ and $P = g'(k)^2 + (a\beta - 1)(y - g(k))g''(k) + a - a\beta g'(k)^2 - a^2\beta$. For β small enough, one can choose a so that S is negative and P is positive. Thus **H'3** of Chapter 4 is fulfilled. Since $\frac{\partial F}{\partial k}(k,y) = (y - g(k))g'(k) - ak + b$ and $\frac{\partial F}{\partial y}(k,y) = -y + g(k) + a\beta y - b\beta$, one may then choose b so that **H2** of Chapter 8 is fulfilled. □

Remark 9.2.1. 1) A stronger version of this theorem may be found in Mitra and Sorger [60]. The C^2 assumption on g is replaced to

Lipschitz continuity.
2) The discount factor β^* associated to this construction is in general quite low. In the case of the logistic map, Sorger [79] finds $\beta^* = \frac{1}{24}$. An associated utility function is :

$$F(k, y) = -16k^4 + 32k^3 - \frac{1}{\beta}k^2 - 16\beta y^2 - 8yk^2 + 8yk + bk - by\beta$$

One sets $a = \frac{1}{\beta} - 16$ in the construction of the theorem.

Further questions. It follows from last section that any Lipschitz continuous function is rationalizable. This result raises the following questions:
Are there continuous functions that cannot occur as optimal policy of some optimal growth problem fulfilling **H1**, **H'2**, **H'3** of Chapter 4, **H2** of Chapter 8.
Are there optimal policies that are not Lipschitz continuous?
The answer to both questions is affirmative. The answer to the first was provided by Sorger [81] and by Mitra and Sorger [60]. For example, the function $g(x) = x^{\frac{1}{3}}$ on $[-1, 1]$ with infinite slope at an interior fixed point of g cannot be the optimal policy function of an optimal growth model.
We would first like to recall that Proposition 9.1.2 provides a sufficient condition for the differentiability of the optimal policy. One may find in Mitra and Sorger [60] examples of optimal policy that have infinite slope at an interior point.

9.3 Appendix

Let

$$F(k, y) = \max_{k^1, k^2, l^1, l^2} f^c(k^1, l^1)$$

under the constraints:

$$k^1 + k^2 \leq k,$$

$$l^1 + l^2 \leq 1,$$
$$0 \leq y \leq f^k(k^2, l^2),$$
$$k^1 \geq 0,\ k^2 \geq 0,\ l^1 \geq 0,\ l^2 \geq 0.$$

At the optimum, the constraints will be binding, i.e.

$$k^1 + k^2 = k,$$

$$l^1 + l^2 = 1,$$
$$0 \leq y = f^k(k^2, l^2).$$

We distinguish two cases.

Case 1 : The functions f^c, f^k are twice continuously differentiable every where.

Since f^k is homogenous of degree one, we have:

$$y = l^2 f^k(\frac{k^2}{l^2}, 1),$$

or equivalently

$$\frac{k^2}{l^2} = \psi(\frac{y}{l^2})$$

where ψ is a convex function, twice continuously differentiable almost everywhere.
The problem becomes:

$$\max_{l^2}(1 - l^2) f^c(\frac{k - l^2\psi(\frac{y}{l^2})}{1 - l^2}, 1).$$

The first order condition is:

$$-f^c(\frac{k^1}{l^1}, 1) + (\frac{k^1}{l^1} - \psi(\frac{y}{l^2}) + \frac{y}{l^2}\psi'(\frac{y}{l^2}))\frac{\partial f^c}{\partial k}(\frac{k^1}{l^1}, 1) = 0. \qquad (9.1)$$

We will denote by l^2 the optimal value of this variable, i.e., the value which satisfies

$$F(k,y) = (1-l^2)f^c\left(\frac{k - l^2\psi(\frac{y}{l^2})}{1-l^2}, 1\right),$$

and by $l^1 = 1 - l^2$ the optimal value of labour in sector 1. It is obvious that:

$$F_1(k,y) = \frac{\partial f^c}{\partial k}\left(\frac{k - l^2\psi(\frac{y}{l^2})}{1-l^2}, 1\right).$$

It is easy to check that

$$F_{12}(k,y) = \frac{1}{l^1}\left(\frac{\partial^2 f^c}{\partial k^2}\right)A$$

where

$$A = \left(\frac{k^1}{l^1} - \psi\left(\frac{y}{l^2}\right) + \frac{y}{l^2}\psi'\left(\frac{y}{l^2}\right)\right)\frac{\partial l^2}{\partial y} - \psi'\left(\frac{y}{l^2}\right).$$

Hence

$$\text{sign } F_{12} = -\text{sign } \left[\left(\frac{k^1}{l^1} - \psi\left(\frac{y}{l^2}\right) + \frac{y}{l^2}\psi'\left(\frac{y}{l^2}\right)\right)\frac{\partial l^2}{\partial y} - \psi'\left(\frac{y}{l^2}\right)\right] \quad (9.2)$$

We will compute $\frac{\partial l^2}{\partial y}$. Using the first order condition, equation 9.1, tedious computations give:

$$\frac{\partial l^2}{\partial y} = \frac{B}{C}$$

with

$$B = \frac{y}{(l^2)^2}\psi''\left(\frac{y}{l^2}\right)\frac{\partial f^c}{\partial k} - \left(\frac{k^1}{l^1} - \psi\left(\frac{y}{l^2}\right) + \frac{y}{l^2}\psi'\left(\frac{y}{l^2}\right)\right)\frac{\psi'(\frac{y}{l^2})}{l^1}\frac{\partial^2 f^c}{\partial k^2},$$

and

$$C = \frac{y^2}{(l^2)^3}\psi''\left(\frac{y}{l^2}\right)\frac{\partial f^c}{\partial k} - \frac{1}{l^1}\left(\frac{k^1}{l^1} - \psi\left(\frac{y}{l^2}\right) + \frac{y}{l^2}\psi'\left(\frac{y}{l^2}\right)\right)^2\frac{\partial^2 f^c}{\partial k^2}.$$

Putting this value of $\frac{\partial l^2}{\partial y}$ in equation 9.2, we obtain:

$$\text{sign } F_{12} = -\text{sign } \left[\left(\frac{k^1}{l^1} - \psi\left(\frac{y}{l^2}\right)\right)\frac{y}{(l^2)^2}\psi''\left(\frac{y}{l^2}\right)\frac{\partial f^c}{\partial k}\right].$$

Since $\psi(\frac{y}{l^2}) = \frac{k^2}{l^2}$ and $\psi''(\frac{y}{l^2}) > 0$, we have

$$\text{sign } F_{12} = \text{sign } (\frac{k^2}{l^2} - \frac{k^1}{l^1}).$$

Case 2 : The function f^c is twice continuously differentiable and $f^k(k,l) = \min\{\frac{k}{\gamma}, \frac{l}{\zeta}\}$.

We have

$$F(k,y) = f^c(k - \gamma y, 1 - \zeta y).$$

It is easy to compute $F_{12}(k,y)$ which equals

$$F_{12}(k,y) = \frac{1}{l^1}\frac{1}{\zeta}(\frac{\partial^2 f^c}{\partial k^2})[\frac{k^2}{l^2} - \frac{k^1}{l^1}].$$

A. Appendix

A.1 Metric Spaces and Normed Spaces

A.1.1 Distances and Metric Spaces

Let E be a set. A *distance* on E is a mapping d of $E \times E$ into $\mathbb{R}$, having the following properties:

(a) $d(x,y) \geq 0$ for any pair $x, \ y$ in E.

(b) $d(x,y) = 0$ if and only if $x = y$.

(c) $d(x,y) = d(y,x)$ for any pair $x, \ y$ in E.

(d) $d(x,z) \leq d(x,y) + d(y,z)$ for any three elements $x, \ y, \ z$ of E.

One can easily check that

$$|d(x,z) - d(y,z)| \leq d(x,y).$$

Let $E = \mathbb{R}^n$. Let $(x = x_1, ..., x_n), \ y = (y_1, ..., y_n)$ be two elements of $\mathbb{R}^n$. The usual Euclidean distance

$$d_1(x,y) = \left(\sum_{i=1}^{n} (x_i - y_i)^2 \right)^{\frac{1}{2}}$$

satisfies the above properties. We can check that the following functions on $\mathbb{R}^n \times \mathbb{R}^n$

$$d_2(x,y) = |x_1 - y_1| + ... + |x_n - y_n|$$

$$d_3(x,y) = \max\{|x_i - y_i| : i = 1, ..., n\}$$

define two other distances on E.

A *metric space* is a set E together with a given distance d on E.

A.1.2 Topology on Metric Spaces

Let E be a metric space with a distance d. Let $a \in E$ and $r > 0$ be a real number. The *open ball* (respectively *closed ball, sphere*) of center a and radius r is the set $B(a,r) = \{x \in E : d(a,x) < r\}$ (resp. $\bar{B}(a,r) = \{x \in E : d(a,x) \leq r\}$, $S(a,r) = \{x \in E : d(a,x) = r\}$).

Open Sets, Closed sets. An *open* set is a subset A of E having the following property: for every $x \in A$, there exists $r > 0$ such that $B(x,r) \subset A$. The empty set is open; the whole space E is open. In particular, any open ball is open. One can verify that the union of any finite or infinite family of open sets is open, and the intersection of a *finite* family of open sets is also open.
The collection $\mathcal{J}$ of open sets of E is called a *topology* on E.
Let d' be another distance on E. It defines another topology $\mathcal{J}'$ on E. We can verify that $\mathcal{J}$ and $\mathcal{J}'$ are *equal* if and only if the following condition is satisfied:
For each $x \in E$ and each open set U in $\mathcal{J}$ containing x, there exists a set U' open in $\mathcal{J}'$ such that $x \in U' \subset U$, and conversely, given U' open in $\mathcal{J}'$ containing x, there exists U open in $\mathcal{J}$ such that $x \in U \subset U'$.
If A is a non-empty subset of E, an *open neighborhood* of A is an open set containing A. A *neighborhood* of A is any set containing an open neighborhood of A. As for the open sets, the intersection of a finite number of neighborhoods of A is a neighborhood of A.
A point x is interior to A if there exists an open ball $B(x,r)$ included in A. The set of all points interior to A is called the *interior* of A and written *int* A.
A subset A is *closed* if its complement in E is open. It is easy to verify that a closed ball is closed and a sphere is closed. The intersection of any finite or infinite family of closed sets is closed. The union of a *finite* number of closed sets is closed.
A *cluster point* of a subset A is a point $x \in E$ such that every open ball centered in x has a non-empty intersection with A. The set of all cluster points of A is called the closure of A and denoted by $\bar{A}$.
We say that $a \in E$ is the limit of a sequence $\{a_n\} \subset E$ if and only if the real sequence $\{d(a, a_n)\}$ converges to zero when n converges to

$+\infty$. We mention that if a sequence $\{a_n\}$ converges to a, then any subsequence $\{a_{n_k}\}$ converges also to a. An element a is a cluster point to $A \subset E$ if and only if there exists a sequence $\{a_n\} \subset A$ which converges to a.

A *Cauchy sequence* is an infinite sequence $\{a_n\}$ such that, for any $\varepsilon > 0$, there exists an integer n_0 such that the relations $p \geq n_0$ and $q \geq n_0$ imply $d(x_p, x_q) < \varepsilon$.

Any convergent sequence is a Cauchy sequence. A metric space E is called *complete* if any Cauchy sequence in E is convergent. The space $\mathbb{R}^n$ endowed with $d_i, i = 1, 2, 3$ defined above is a complete metric space.

Assume $E = \mathbb{R}$. Let $d(x, y) = |x - y|$ for any pair x, y of E. Let $\{a_n\}$ be a sequence of real numbers. We define:

$$\limsup_{n \to +\infty} a_n = \lim_{T \to +\infty} \sup\{a_n : n \geq T\},$$

$$\liminf_{n \to +\infty} a_n = \lim_{T \to +\infty} \inf\{a_n : n \geq T\}.$$

These limits exist in $\mathbb{R} \cup \{-\infty\} \cup \{+\infty\}$, and $\limsup_{n \to +\infty} a_n \geq \liminf_{n \to +\infty} a_n$. When these limits are equal, the sequence $\{a_n\}$ has a limit and:

$$\lim_{n \to +\infty} a_n = \liminf_{n \to +\infty} a_n = \limsup_{n \to +\infty} a_n.$$

Let us mention the following useful inequalities :

$$\liminf_{n \to +\infty} a_n + \liminf_{n \to +\infty} b_n \leq \liminf_{n \to +\infty} (a_n + b_n)$$

$$\leq \limsup_{n \to +\infty} (a_n + b_n) \leq \limsup_{n \to +\infty} a_n + \limsup_{n \to +\infty} b_n$$

Continuous Mappings. Let E and E' be two metric spaces, d, d' the distances on E and E'. A mapping f of E into E' is *continuous at* $a \in E$ if, for every neighborhood V' of $f(a)$ in E', there exists a neighborhood V of a in E such that $f(V) \subset V'$. The mapping f is said to be *continuous* if it is continuous at *every point* of E.

The mapping f is continuous at a if and only if $f^{-1}(V')$ is a neighborhood of a for any neighborhood V' of $f(a)$. One can also characterize the continuity of f as follows: for any sequence $\{a_n\} \subset E$

converging to a, the sequence $\{f(a_n)\}$ converges to $f(a)$. The following statements are equivalent:
(i) f is continuous;
(ii) for every open set $A' \subset E'$, $f^{-1}(A')$ is open in E.
(iii) for every closed set $A' \subset E'$, $f^{-1}(A')$ is closed in E.

Monotonic Functions. Let f be a function from $\mathbb{R}$ into $\mathbb{R}$. We say that f is *increasing* if, $\forall x \in \mathbb{R}, \forall y \in \mathbb{R},\ y > x \Rightarrow f(y) > f(x)$. It is *decreasing* if, $\forall x \in \mathbb{R}, \forall y \in \mathbb{R},\ y > x \Rightarrow f(y) < f(x)$. It is *non-decreasing* (resp. *non-increasing*) if, $y > x \Rightarrow f(y) \geq f(x)$ (resp. $f(y) \leq f(x)$).
A function f is *monotonic* if either it is non-increasing or non-decreasing.
We have the following property: if f is monotonic then it is continuous almost everywhere. More precisely, the set of elements where f is not continuous is finite or countable (a set A is *countable* if there exists a bijection of A into the set $\mathbb{N}$ of integers).

Contraction Mappings.

Theorem A.1.1 (Contraction Mapping Theorem). *Let E be complete metric space with distance d. Let A be a closed, non-empty subset of E. Let f be a contraction of A, i.e., a mapping from A into A which satisfies:*

$$\exists k \in]0,1[,\ \forall x \in A,\ \forall y \in A,\ d(f(x), f(y)) \leq k d(x,y).$$

Then there exists a unique fixed-point of f, i.e., a unique $\bar{x}$ which satisfies $\bar{x} = f(\bar{x})$. Moreover, $\bar{x} = \lim_{n\to+\infty} f^n(x)$, $\forall x \in A$. Here, f^n denotes the n-iterated of f.

Upper Semi-continuous Functions. Let f be a function from a complete metric space E into R. We say that f is *upper semi-continuous* if, for any $\alpha \in \mathbb{R}$, the set $\{x : f(x) \geq \alpha\}$ is closed in E. An equivalent definition is as follows. The function f is *upper semi-continuous* if, for any $x \in E$, for any sequence $\{x_n\} \subset E$ converging to x, we have $\limsup_{n\to+\infty} f(x_n) \leq f(x)$. Obviously, a continuous function is upper semi-continuous.
It is easy to show that the limit of any a sequence of decreasing, upper semi- continuous functions on E is upper semi-continuous on E.

A.1.3 Compact Spaces, Compact Sets

Let E be a metric space. We say that the collection of open sets of E, $\{U_i\}_{i\in I}$ is a *covering* of E if $E \subset \cup_{i\in I} U_i$. The metric space E is *compact* if, for every covering $\{U_i\}_{i\in I}$ of E, there exists a *finite* subcovering of E.
The following statements are equivalent:
(i) E is compact
(ii) Any infinite sequence in E has at least a cluster point.
In a compact metric space E, any infinite sequence which has a unique cluster point is convergent (and obviously, it converges to this cluster point).
A subset A of a metric space E is *compact* if for any covering of A there exists a *finite* subcovering.
Any compact set of a metric space is closed and any closed subset of a compact set is compact.
If A is compact in a metric space E, and if f is a continuous mapping from E into a metric space E', then $f(A)$ is a compact set of E'.
The following property is very important. Let f be an upper semi-continuous function on A, non-empty subset of a metric space E. If A is compact, then f has a maximum on A. Obviously, this property also holds for continuous function. Moreover, if f is continuous and A is compact, then f has also a minimum on A.

A.1.4 Normed Spaces

Let E be a *vector space.* A *norm* in E is a function (usually written $x \longrightarrow \|x\|$) of E into $\mathbb{R}$, having the following properties:
(1) $\|x\| \geq 0, \ \forall x \in E$.
(2) $\|x\| = 0 \ \Leftrightarrow x = 0$.
(3) $\|\lambda x\| = |\lambda| \|x\|, \ \forall x \in E, \forall \lambda \in \mathbb{R}$.
(4) $\|x + y\| \leq \|x\| + \|y\|, \ \forall x \in E, \ \forall y \in E$.
A vector space E endowed with a norm is called *normed space.*
Let $d(x, y) = \|x - y\|$. It is easy to verify that d defines a distance on E.
E is called *Banach space* if, with this distance d, it is a complete metric space.

Some examples of Banach spaces:

$\mathbb{R}^n$ is a Banach space with the following norms:

(i) $\|x\| = \max\{|x_i| : i = 1, ..., n\}$ (sup-norm).

(ii) $\|x\| = (\sum_{i=1}^n {x_i}^2)^{\frac{1}{2}}$ (Euclidean norm).

(iii) $\|x\| = |x_1| + ... + |x_n|$.

Let $\mathcal{B}$ be the set of bounded functions from a metric space E into $\mathbb{R}$. Define

$$\forall f \in \mathcal{B}, \ \|f\| = \sup\{|f(x)| : x \in E\}.$$

$\mathcal{B}$ is a Banach space with this norm.
The set of continuous, bounded functions from a metric space into $\mathbb{R}$ is also a Banach space with the previous norm.
The following property is very important: in $\mathbb{R}^n$, a set is compact if and only if it is norm bounded (i.e., it is included in a ball of $\mathbb{R}^n$) and closed. This property does not hold for infinite dimensional spaces.
The following result is also very important. Let $\{f_n\}$ be a sequence of *continuous* mappings from a compact set X of a metric space E into a normed space F. Assume that $\{f_n\}$ converges *uniformly* to a mapping f from X into F, i.e.

$$\forall \epsilon > 0, \exists N_0, \text{such that } \forall n \geq N_0, \forall x \in X, \|f_n(x) - f(x)\| \leq \epsilon.$$

Then, the mapping f is *continuous* from X into F.

A.1.5 Product Topology

Let $\mathbb{R}^n$ be endowed with some norm $\|\|$. Let $(\mathbb{R}^n)^{+\infty}$ denote the spaces of infinite sequences $\mathbf{x} = (x_0, x_1, ..., x_n, ...)$ endowed with the distance

$$\forall \mathbf{x} \in (\mathbb{R}^n)^{+\infty}, \forall \mathbf{y} \in (\mathbb{R}^n)^{+\infty}, \ d(\mathbf{x}, \mathbf{y}) = \sum_{i=0}^{+\infty} \frac{1}{2^i} \left(\frac{\|x_i - y_i\|}{1 + \|x_i - y_i\|} \right).$$

The topology on $(\mathbb{R}^n)^{+\infty}$ associated with this distance is called *product topology*. Observe that if $(\mathbf{x^0}, ..., \mathbf{x^n})$ is a sequence of elements of $(\mathbb{R}^n)^{+\infty}$ which converges to some $\mathbf{x} \in (\mathbb{R}^n)^{+\infty}$, then $\forall i, x_i^n \to x_i$ when $n \to +\infty$.

For this topology, the product of an infinite family of compact sets of $\mathbb{R}^n$ is compact, i.e., if $(A_0, A_1, ..., A_n, ...)$ is an infinite sequence of compact sets of $\mathbb{R}^n$, then the set $\Pi_{i=0}^{+\infty} A_i$ is compact for the product topology.

A.2 Concave Functions

A set A of $\mathbb{R}^n$ is convex if $\forall x \in A, \forall y \in A, \forall \lambda \in [0,1],\ \lambda x + (1-\lambda) y \in A$.
Let f be a function from a convex set A into $\mathbb{R} \cup \{-\infty\}$. It is *concave* if

$$\forall x \in A, \forall y \in A, \forall \lambda \in [0,1],\ f(\lambda x + (1-\lambda) y) \geq \lambda f(x) + (1-\lambda) f(y).$$

The domain of a concave function f , denoted by $\mathrm{dom}(f)$, is the set of elements of A such that $f(x) > -\infty, \forall x \in \mathrm{dom}(f)$. The domain of f is convex.
It is important to notice that a concave function is *continuous* in the interior of its domain.
We have the *Jensen Inequality* for a concave function:
A function f is concave on a convex set A if, and only if, for any integer $p \geq 2$, for any p elements of A, $x_1, ..., x_p$, for any positive numbers $\lambda_1, ..., \lambda_p$, the sum of which equals 1, one has

$$f(\lambda_1 x_1 + ... + \lambda_p x_p) \geq \lambda_1 f(x_1) + ... + \lambda_p f(x_p).$$

More generally, if $\{\lambda_i\}$ is a sequence of non-negative numbers such that $\sum_{i=1}^{+\infty} \lambda_i = 1$, and if $\{x_i\}$ is a sequence of elements in A, then

$$f(\sum_{i=1}^{+\infty} \lambda_i x_i) \geq \sum_{i=1}^{+\infty} \lambda_i f(x_i).$$

Obviously, the limit of a sequence of concave functions is concave.
A function f from a convex set A into $\mathbb{R}$ is *strictly concave* if

$$\forall \lambda \in]0,1[, \forall (x,y) \in A \times A, x \neq y,$$

$$f(\lambda x + (1-\lambda) y) > \lambda f(x) + (1-\lambda) f(y).$$

Assume that the concave function f has a maximum on a convex set A. If f is strictly concave, then the maximum is unique.

A.2.1 Subdifferentiability, Differentiability of Concave Functions

Let f be a concave function on $\mathbb{R}^n$. The element $x^* \in \mathbb{R}^n$ is called *subgradient* of f at x if $f(x) - f(z) \geq x^*.(x - z), \forall z \in \mathbb{R}^n$. The set of subgradients of f at x is called *subdifferential* of f at x and is noted $\partial f(x)$. If $\partial f(x)$ is non-empty, we say that f is *subdifferentiable* at x. Observe that, $\partial f(x)$ is closed and convex.
If $x \notin \text{dom}(f)$, then $\partial f(x)$ is empty. If $x \in \textit{int}\,\text{dom}(f)$, then $\partial f(x)$ is compact, non-empty.
We now prove that any concave function f is bounded above by an affine function. For simplicity, we assume that $\textit{int}\,\text{dom}(f)$ is non-empty. Take some $x_0 \in \textit{int}\,\text{dom}(f)$. The subdifferential $\partial f(x_0)$ is non-empty. Let $p \in \partial f(x_0)$. We then have

$$\forall x, \ f(x_0) - f(x) \geq p \cdot (x_0 - x),$$

hence,

$$\forall x, \ f(x) \leq p \cdot x + f(x_0) - p \cdot x_0.$$

In particular, if f is non-negative and concave, then

$$\forall x, \ |f(x)| \leq \|p\| \|x\| + |f(x_0)| + |p \cdot x_0|.$$

If f is strictly concave , then

$$\forall x^* \in \partial f(x), \forall z \neq x, f(x) - f(z) > x^*.(x - z).$$

A concave function f on $\mathbb{R}^n$ is differentiable at x if, and only if, $\partial f(x) = \{x^*\}$. In that case, the gradient of f, $f'(x) = x^*$.
Let f be a concave function from an interval $[a, b] \subset \mathbb{R}$ into $\mathbb{R}$. Then: f has left and right derivatives, f'_-, f'_+ in $]a, b[$ and $\partial f(x) = [f'_+(x), f'_-(x)]$. More precisely, if $a < x < x' < b$, then

$$f'_-(x) \geq f'_+(x) \geq \frac{f(x') - f(x)}{x' - x} \geq f'_-(x') \geq f'_+(x').$$

We deduce from this inequality that, if f is concave, differentiable, then f' is a non-increasing function.

The following results are very important. Assume that f is a function, twice differentiable on an open convex, non-empty set A of $\mathbb{R}^n$.
Then f is concave if, and only if, for any $a_0 \in A$, the Hessian matrix $\left(\frac{\partial^2 f}{\partial x_i \partial x_j}\right)(a_0)$ is non-positive, i.e., for any $(h_1, ..., h_n) \in \mathbb{R}^n$, we have $\sum_{i,j} h_i h_j \frac{\partial^2 f}{\partial x_i \partial x_j}(a_0) \leq 0$.
A sufficient condition for a function f is *strictly* concave is that the Hessian matrix $\left(\frac{\partial^2 f}{\partial x_i \partial x_j}\right)(a_0)$ is negative definite, i.e.,

$$\forall (h_1, ..., h_n) \in \mathbb{R}^n, \sum_{i,j} h_i h_j \frac{\partial^2 f}{\partial x_i \partial x_j}(a_0) < 0,$$

if $h \neq 0$.

A.3 Negative Matrix

A $n \times n$-matrix A is called *non-positive* matrix if, for any vector $h = (h_1, ..., h_n) \in \mathbb{R}^n$, the real number ${}^t hAh = \sum_{i,j} h_i h_i A_{ij} \leq 0$.
The matrix A is *negative* if ${}^t hAh = \sum_{i,j} h_i h_i A_{ij} < 0, \ \forall h \in \mathbb{R}^n, h \neq 0$.
A symmetric square n-matrix A is *negative definite* if ${}^t hAh = \sum_{i,j} h_i h_i A_{ij} < 0, \ \forall h \in \mathbb{R}^n, h \neq 0$.
In this case, the matrix A is diagonalisable. Moreover, there exists a matrix P such that $A = {}^t P \begin{pmatrix} \lambda_1 & & \\ & \dots & \\ & & \lambda_n \end{pmatrix} P$ where ${}^t P$ is the transposed matrix of P, and $\lambda_i < 0, \forall i = 1, ..., n$.The numbers λ_i are the eigenvalues of A. The matrix P has this property: if $x = (x_1, ..., x_n) \in \mathbb{R}^n$, and if

$$\begin{pmatrix} x' \\ y' \end{pmatrix} = P \begin{pmatrix} x \\ y \end{pmatrix}$$

then $x_1^2 + ... + x_n^2 = {x'_1}^2 + ... + {x'_n}^2$.
In particular, we have:

$$(\min_i \lambda_i)\|x\|^2 \leq {}^t xAx \leq (\max_i \lambda_i)\|x\|^2.$$

A.4 The Implicit Functions Theorem

Theorem A.4.1. *Let f be a p times continuously differentiable from $U \times V$ into $\mathbb{R}^n$, where U is an open, non-empty set of $\mathbb{R}^m$ and V is an open, non-empty set of $\mathbb{R}^n$. Suppose that $f(a,b) = 0$ and that the partial derivative of f, with respect to the second variable, at (a,b), $f_2(a,b)$, is invertible. Then there exists an open set $U_1 \subset U$, containing a, an open set $V_1 \subset V$ containing b, an open set W containing 0 and a function $g : U_1 \times W \to V_1$ such that :*

$$\forall x \in U_1, \forall y \in V_1, \forall z \in W, f(x,y) = z \Leftrightarrow y = g(x,z).$$

Moreover, g is p times continuously differentiable on $U_1 \times W$. In particular,

$$(x,y) \in U_1 \times W, f(x,y) = 0 \Leftrightarrow y = \phi(x),$$

where ϕ is p times continuously differentiable on U_1.

A.5 Correspondences, The Maximum Theorem

Let $X \subset \mathbb{R}^l$, $Y \subset \mathbb{R}^m$. A *correspondence* Γ from X into Y is a mapping from X into the set of subsets of Y. The graph of Γ, noted graph Γ, is the set

$$\text{graph}\, \Gamma = \{(x,y) \in X \times Y : y \in \Gamma(x)\}.$$

A correspondence $\Gamma : X \to Y$ is *lower semi-continuous* at point x if for any $y \in \Gamma(x)$ and for any sequence $\{x_n\} \subset X$ converging to x, there exist m and a sequence $\{y_n\}$ with $y_n \in \Gamma(x_n), \forall n \geq m$, such that $\{y_n\}$ converges to y when n converges to $+\infty$.
Γ is lower semi-continuous on X if it is lower semi-continuous everywhere on X.
A correspondence $\Gamma : X \to Y$ is *closed* if its graph is closed.
A correspondence $\Gamma : X \to Y$ is *upper semi-continuous* at point x if
(i) $\Gamma(x)$ is compact, non-empty,

(ii) For any sequence $\{x_n\}$ converging to x, for any sequence $\{y_n\}$ with $y_n \in \Gamma(x_n), \forall n$, there exists subsequence $\{y_{n_k}\}$ converging to $y \in \Gamma(x)$.
Γ is upper semi-continuous on X if it is upper semi-continuous everywhere on X.
A correspondence Γ is *continuous* at x, if it is upper semi-continuous and lower semi-continuous at x. It is continuous if it is continuous everywhere on X.

Remark A.5.1. (i) When Γ is a mapping, i.e. single-valued correspondence, it is continuous if, and only if, either it is upper semi-continuous or it is lower semi-continuous.
(ii) If X is compact, then Γ is upper semi-continuous on X if, and only if, it is closed.
(iii) If Γ is upper semi-continuous on X and if $K \subset X$ is compact, then $\Gamma(K)$ is compact in Y.

A.5.1 Examples

(1) Let $\Gamma : \mathbb{R}_+ \to \mathbb{R}_+$ be defined by $\Gamma(x) = [0, f(x)]$ where f is a continuous mapping, then Γ is continuous.
(2) Let $\Gamma : \mathbb{R} \to \mathbb{R}$ be defined by:

$$\Gamma(x) = \{-1, +1\}, \text{ if } x > 0,$$

$$\Gamma(0) = [-1, +1],$$
$$\Gamma(x) = 0, \text{ if } x < 0.$$

Γ is upper semi-continuous, but not lower semi-continuous at 0.
Let $\Gamma : \mathbb{R} \to \mathbb{R}$ be defined by $\Gamma(x) = [-1, +1]$ if $x > 0$, and $\Gamma(x) = \{0\}$ if $x \leq 0$. The correspondence Γ is lower semi-continuous at 0 but not upper semi-continuous at 0.

Theorem A.5.1 (The Maximum Theorem). *Let $Y \subset \mathbb{R}^l$, $X \subset \mathbb{R}^m$, $\phi : Y \to \mathbb{R}$ be a continuous function, and $\Gamma : X \to Y$ be a continuous correspondence with non-empty values. Then the function h defined by*

$$h(x) = \max\{\phi(y) : y \in \Gamma(x)\}$$

is a continuous function.

The correspondence G defined by:

$$G(x) = \{y : y \in \Gamma(x), \phi(y) = h(x)\}$$

is an upper semi-continuous correspondence on X.

From this theorem, we obtain two important corollaries.

Corollary A.5.1. *Let Γ be a continuous correspondence from $X \subset \mathbb{R}^l$ into $Y \subset \mathbb{R}^m$ and let $f : graph\Gamma \to \mathbb{R}$ be a continuous function. Then the function h defined by:*

$$h(x) = \max\{f(x, y) : y \in \Gamma(x)\}$$

is a continuous function on X. The correspondence G defined by:

$$G(x) = \{y : y \in \Gamma(x), f(x, y) = h(x)\}$$

is upper semi-continuous.

Proof. Let $z = (x, y) \in graph\Gamma$. Let $\gamma : X \to X \times Y$ be the correspondence defined by $\gamma(x) = (x, \Gamma(x))$. The problem becomes $\max\{f(z) : z \in \gamma(x)\}$. Obviously, the correspondence γ is continuous. Apply the Maximum Theorem. □

Corollary A.5.2. *If $\Gamma : X \to Y$ is a continuous correspondence with* **convex** *values and if $f : graph\Gamma \to \mathbb{R}$ is a continuous function such that $f(x, .)$ is* **strictly concave** *for every x, then the correspondence G defined by*

$$G(x) = \{y : y \in \Gamma(x), f(x, y) = h(x)\}$$

is a continuous **mapping** *(i.e. single-valued correspondence).*

Proof. Because of the strict concavity of f with respect to the second variable, the solution is unique and hence, G is single-valued. □

We now give a weak form of the Maximum Theorem.

Theorem A.5.2. *If $f(x, y)$ is an upper semi-continuous real function on $X \times Y$, and if Γ is an upper semi-continuous correspondence from X into Y, then the function h defined by:*

$$h(x) = \max\{f(x, y) : y \in \Gamma(x)\}$$

is upper semi-continuous.

Bibliography

[1] P. Aghion and P. Howitt. *Endogenous growth theory.* 2d. ed. THe MIT Press, 1998.

[2] C.D. Aliprantis, K.C. Border, and O. Burkinshaw. New proof of the existence of equilibrium in a single-sector growth model. *Macroeconomic dynamics* **1**, pages 669–679, 1997.

[3] F. Alvarez and N. Stokey. Dynamic programming with homogeneous functions. *Journal of economic theory* **82**, pages 167–189, 1998.

[4] R. Amir. Sensivity analysis in multisector optimal economic models. *Journal of mathematical economics* **25**, pages 123–141, 1996.

[5] K.J. Arrow and M.D. Intriligator, editors. *Handbook of mathematical economics, Vol. 3.* North-Holland, Amsterdam, Holland, 1986.

[6] P. Askenazy and C. Le Van. A model of optimal growth strategy. *Journal of economic theory* **85**, pages 24–51, 1999.

[7] K. Askri and C. Le Van. Differentiability of the value function of nonclassical optimal growth models. *Journal of optimization theory and applications* **97**, pages 591–604, 1998.

[8] R.J. Barro and X. Sala-I-Martin. *Economic growth.* McGraw-Hill, Inc., 1995.

[9] R.A. Becker and J. H. Boyd III. *Capital Theory, Equilibrium Analysis and Recursive Utility.* Blackwell Publishers, 1995.

[10] R. Bellman. *Dynamic Programming.* Princeton:Princeton University Press, 1957.

[11] J. Benhabib and K. Nishimura. Competitive equilibrium cycles. *Journal of economic theory* **35**, pages 284–306, 1985.

[12] L.M. Benveniste and J.A. Scheinkman. On the differentiability of the value function in dynamic models of economies. *Econometrica* **47**, pages 727–732, 1979.

[13] C. Berge. *Espaces Topologiques.* Paris: Dunod, English translation: Topological Spaces, Edinburgh London: Oliver and Boyd, 1963, 1959.

[14] T.F. Bewley. Existence of equilibria in economies with infinitely many commodities. *Journal of economic theory* **4**, pages 514–540, 1972.

[15] T.F. Bewley. An integration of equilibrium theory and turnpike theory. *Journal of mathematical economics* **10**, pages 233–267, 1982.

[16] M. Boldrin and R. Deneckere. Sources of complex dynamics in two sector growth models. *Journal of economic dynamics and control* **14**, pages 627–653, 1990.

[17] M. Boldrin and L. Montrucchio. On the indeterminacy of capital accumulation paths. *Journal of economic theory* **40**, pages 26–39, 1986.

[18] J.H. Boyd III. Recursive utility and the ramsey model. *Journal of economic theory* **50**, pages 326–345, 1990.

[19] W.A. Brock. On existence of weakly maximal programmes in a multi-sector economy. *Review of economic studies* **37**, pages 275–280, 1970.

[20] H. Cartan. *Calcul Différentiel.* Hermann Paris, Collection Méthodes, 1967.

[21] D. Cass. Optimum growth in an aggregative model of accumulation. *Review of economic studies* **32**, pages 233–240, 1965.

[22] G. Choquet. *Topologie.* Masson et Cie, 1969.

[23] C.W. Clark. Economically optimal policies for the utilization of biologically renewable resources. *Mat. Biosci.* **17**, pages 245–268, 1971.

[24] F.H. Clarke. *Optimization and nonsmooth analysis.* A Wiley-interscience publication, 1983.

[25] R.A. Dana and C. Le Van. On the bellman equation of the overtaking criterion. *Journal of optimization theory and applications* **67**, pages 587–600, 1990.

[26] R.A. Dana and C. Le Van. Equilibria of a stationary economy with recursive preferences. *Journal of optimization theory and applications* **71**, pages 289–313, 1991.

[27] R.A. Dana and C. Le Van. On the bellman equation of the overtaking criterion: addendum. *Journal of optimization theory and applications* **78**, pages 605–612, 1993.

[28] G. Debreu. Excess demand functions. *Journal of mathematical economics* **1**, pages 15–21, 1973.

[29] W.D. Dechert and K. Nishimura. A complete characterization of optimal growth paths in an aggregated model with a non-concave production function. *Journal of economic theory* **31**, pages 332–354, 1983.

[30] R. Deneckere and S. Pelikan. Competitive chaos. *Journal of economic theory* **40**, pages 13–25, 1986.

[31] J. Dieudonne. *Foundations of Modern Analysis.* Academic Press, 1969.

[32] C.-H. Dimaria and C. Le Van. Debt, corruption, r&d and growth in developing countries. *Macroeconomic Dynamics* , 2002.

[33] E. Domar. Capital expansion, rate of growth, and employment. *Econometrica* **14**, pages 137–147, 1946.

[34] E. Domar. Expansion and employment. *American economic review* **37**, pages 34–55, 1947.

[35] J. Duran. On dynamic programming with unbounded returns. *Economic theory* **15**, pages 339–352, 2000.

[36] J. Duran and C. Le Van. Simple proof of existence of equilibrium in a one-sector model with bounded or unbounded returns from below. *Macroeconomic dynamics* , 2002.

[37] M. Florenzano, C. Le Van, and P. Gourdel. *Finite Dimensional Convexity and Optimization.* Springer, 2001.

[38] M. Frankel. The production function in allocation and growth: a synthesis. *American economic review* **55/3**, pages 296–319, 1962.

[39] D. Gale. On optimal development in a multi-sector economy. *Review of economic studies* **34**, pages 1–18, 1967.

[40] J.M. Grandmont. *Periodic and aperiodic in discrete one-dimensional systems.* In W. and Mas-Colell [84], 1986.

[41] I. Hadji and C. Le Van. Convergence of equilibria in an intertemporal general equilibrium model. *Journal of economic dynamics and control* **18**, pages 381–396, 1994.

[42] R.F. Harrod. An essay in dynamic theory. *Economic Journal,* **49**, pages 14–33, 1939.

[43] R.F. Harrod. *Towards a dynamic economics.* London: Macmillan, 1948.

[44] T.C. Koopmans. On the concept of optimal economic growth. *Pontificiae academiae scientarum varia* **28**, pages 225–300, 1965.

[45] S. Lang. *Analysis II.* Addison-Wesley Publishing Company, Inc., 1969.

[46] A. Lasota and J.A. Yorke. On the existence of invariant measures for piecewise monotonic transformations. *AMS* **183**, pages 481–485, 1973.

[47] C. Le Van and L. Morhaim. Optimal growth models with bounded or unbounded returns: a unifying approach. *Journal of economic theory* , 2002.

[48] C. Le Van and Y. Vailakis. Recursive utility and optimal growth with bounded or unbounded returns. *DP CORE* , 2002.

[49] T. Li and J.A. Yorke. Period three implies chaos. *American Mathematical Monthly* **82**, pages 985–992, 1975.

[50] R. Lucas. On the mechanics of economic development. *Journal of monetary economics* **22**, pages 3–42, 1988.

[51] M. Majumdar and T. Mitra. On optimal exploitation of a renewable resource in a non-convex environment and the minimum safe standard of conservation. *W.P. No 223, Cornell University*, 1980.

[52] M. Majumdar and T. Mitra. Intertemporal allocation with a non-convex technology: the aggregative framework. *Journal of economic theory* **27**, pages 101–136, 1982.

[53] R. Mantel. On the characterization of aggregate excess demand. *Journal of economic theory* **7**, pages 348–353, 1974.

[54] T. Maruyama and Takahashi, editors. *Nonlinear dynamics and convex analysis in economic theory.* Springer-Verlag, Berlin, 1986.

[55] L.W. McKenzie. Turnpike theory, discounted utility, and the von-neumann facet. *Journal of economic economic theory* **30**, pages 330–352, 1983.

[56] L.W. McKenzie. *Optimal economic growth, turnpike theorems, and comparative dynamics*, pages 1281–1358. In Arrow and Intriligator [5], 1986.

[57] T. Mitra. Sensitivity of optimal programs with respect to changes in target stocks: the case of irreversible investment. *Journal of economic theory* **29**, pages 172–184, 1983.

[58] T. Mitra. An exact discount factor restriction for period-three cycles in dynamic otimization models. *Journal of economic theory* **69**, pages 281–305, 1996.

[59] T. Mitra and D. Ray. Efficient and optimal programs when investment is irreversible. *Journal of economics* , 1982.

[60] T. Mitra and G. Sorger. Rationalizing policy functions by dynamic optimization. *Econometrica* **67**, pages 375–392, 1999.

[61] L. Montrucchio. Thompson metric, contraction property and differentiability of policy functions. *Journal of economic behaviour&organisation* **33**, pages 449–466, 1998.

[62] K. Nishimura and G. Sorger. Optimal cycles and chaos; a survey. *Studies in Nonlinear Dynamics and Econometrics* **1**, pages 11–28, 1996.

[63] K. Nishimura and M. Yano. *Non-linearity and business cycles in a two-sector equilibrium model: an example with Cobb-Douglas production functions*, pages 231–245. In Maruyama and Takahashi [54], 1986.

[64] K. Nishimura and M. Yano. On the least upper bound of discount factors that are compatible with optimal period-three cycles. *Journal of economic theory* **69**, pages 306–333, 1996.

[65] Stokey N.L. and R. Lucas with E. Prescott. *Recursive Methods in Economic Dynamics.* Harvard University Press, 1989.

[66] F. Ramsey. A mathematical theory of saving. *Economic journal* **38**, pages 543–559, 1928.

[67] A.W. Roberts and D.E. Varberg. *Convex Functions.* New York San Francisco London: Academic Press, 1973.

[68] P. Romer. Increasing returns and long-run growth. *Journal of political economy* **94/5**, pages 1002–1037, 1986.

[69] P. Romer. Endogenous technological change. *Journal of political economy* **98/5**, pages s71–s102, 1990.

[70] Rockafellar R.T. *Convex Analysis.* 2d. ed. Princeton: Princeton University Press, 1972.

[71] W. Rudin. *Functional Analysis.* 2d. ed. McGraw-Hill International Editions, 1991.

[72] P.A. Samuelson. Optimality of profit, incluning prices under ideal planning. *Proceedings of the National Academy of Sciences of the United States of America* **70**, pages 2109–2111, 1973.

[73] P.A. Samuelson and R. Solow. A complete capital model involving heterogeneous capital goods. *Quarterly journal of economics* **70**, pages 537–562, 1956.

[74] M.S. Santos. Smoothness of the policy function in discrete time economic models. *Econometrica* **59**, pages 1365–1382, 1991.

[75] A.N. Sarkovski. Coexistence of cycles of a continuous map of a line into itself. *Ukrainian mathematical journal* **16**, pages 61–71, 1964.

[76] J.A. Scheinkman. On optimal steady states of n-sector growth models when utility is discounted. *Journal of economic theory* **12**, pages 11–30, 1976.

[77] R. Solow. A contribution to the theory of economic growth. *Quarterly journal of economics* **70/1**, pages 65–94, 1956.

[78] H. Sonnenschein. Do walras identity and continuity characterize the class of community excess demand functions? *Journal of economic theory* **6**, pages 345–354, 1973.

[79] G. Sorger. *Minimum impatience theorems for recursive economic models.* Springer-Verlag, 1992.

[80] G. Sorger. Period three implies heavy discounting. *Mathematics of operations research* **19**, pages 1007–1022, 1994.

[81] G. Sorger. On the sensitivity of optimal growth paths. *Journal of mathematical economics* **24**, pages 206–233, 1995.

[82] T. Swann. Economic growth and capital accumulation. *Economic Record* **32**, pages 334–361, 1956.

[83] C.C. von Weizscker. Existence of optimal programmes of accumulation for an infinite time horizon. *Review of economic studies* **32**, pages 85–104, 1965.

[84] Hildenbrand W. and A. Mas-Colell, editors. *Contributions to mathematical economics.* North-Holland, Amsterdam, Holland, 1986.

[85] M.L. Weitzman. Duality theory for infinite horizon convex models. *Management Science* **19**, pages 783–789, 1973.

Index

Dynamic Modeling and Econometrics in Economics and Finance

1. P. Rothman: *Nonlinear Time Series Analysis of Economic and Financial Data*. 1999
ISBN 0-7923-8379-6
2. D.M. Patterson and Richard A. Ashley: *A Nonlinear Time Series Workshop*. 1999
ISBN 0-7923-8674-4
3. A. Mele and F. Fornari: *Stochastic Volatility in Financial Markets*. 2000
ISBN 0-7923-7842-3
4. I. Klein and S. Mittnik (eds.): *Contributions to Modern Econometrics*. 2002
ISBN 1-4020-7334-8
5. C. Le Van and R.-A. Dana: *Dynamic Programming in Economics*. 2003
ISBN 1-4020-7409-3

KLUWER ACADEMIC PUBLISHERS – DORDRECHT / BOSTON / LONDON

FSC
www.fsc.org
MIX
Papier aus verantwortungsvollen Quellen
Paper from responsible sources
FSC® C105338

If you have any concerns about our products,
you can contact us on
ProductSafety@springernature.com

In case Publisher is established outside the EU,
the EU authorized representative is:
Springer Nature Customer Service Center GmbH
Europaplatz 3, 69115 Heidelberg, Germany

Printed by Libri Plureos GmbH
in Hamburg, Germany